BOOK ORDER INFORMATION
REPAIRMASTER
P.O. BOX 649
WEST JORDAN, UT 84084
BANK CARD ORDERS 1-800-347-5163

VOLUME 9 5556 ISBN 1-56302-024-6

32 VCLUME SET ISBN 1-56302-105-6

Copyright 1981

MASTER

Editor/Tech-Writer .... Barnee Schollnick

REPAIR-MASTER for..
WHIRLPOOL
AUTOMATIC
DISHWASHER

PRINTED IN U.S.A.

VOLUME 9 5556 ISBN 1-56302-024-6

32 VOLUME SET ISBN 1-56302-105-6

We have used all possible care to assure the accuracy of the information contained in this book. However, the publisher assumes no liability for any errors, omissions or any defects whatsoever in the diagrams and/or instructions or for any damage or injury resulting from utilization of said diagrams and or instructions.

FOREWORD

This Repair Master contains information and service procedures to assist the service technician in correcting conditions that are not always obvious.

A thorough knowledge of the functional operation of the many component parts used on appliances is important to the serviceman, if he is to make a proper diagnosis when a malfunction of any part occurs.

We have used many representative illustrations, diagrams and photographs to portray more clearly these various components for a better over-all understanding of their use and operation.

IMPORTANT SAFETY NOTICE

You should be aware that all major appliances are complex electromechanical devices. Master Publication's REPAIR MASTER® Service Publications are intended for use by individuals possessing adequate backgrounds of electronic, electrical and mechanical experience. Any attempt to repair a major appliance may result in personal injury and property damage. Master Publications cannot be responsible for the interpretation of its service publications, nor can it assume any libility in connection with their use.

SAFE SERVICING PRACTICES

To preclude the possibility of resultant personal injury in the form of electrical shock, cuts, abrasions or burns, etc., that can occur spontaneously to the individual while attempting to repair or service the appliance; or may occur at a later time to any individual in the household who may come in contact with the appliance, Safe Servicing Practices must be observed. Also property damage, resulting from fire, flood, etc., can occur immediately or at a later time as a result of attempting to repair or service — unless safe service practices are observed.

The following are examples, but without limitation, of such safe practices:

1. Before servicing, always disconnect the source of electrical power to the appliance by removing the product's electrical plug from the wall receptacle, or by removing the fuse or tripping the circuit breaker to OFF in the branch circuit servicing the product.

NOTE: If a specific diagnostic check requires electrical power to be applied such as for a voltage or amperage measurements, reconnect electrical power only for time required for specific check, and disconnect power immediately thereafter. During any such check, ensure no other conductive parts, panels or yourself come into contact with any exposed current carrying metal parts.

2. Never bypass or interfere with the proper operation of any feature, part, or device engineered into the appliance.

3. If a replacement part is required, use the specified manufacturers part, or an equivalent which will provide comparable performance.

4. Before reconnecting the electrical power service to the appliance — be sure that:

    a. All electrical connections within the appliance are correctly and securely connected.
    b. All electrical harness leads are properly dressed and secured away from sharp edges, high-temperature components such as resistors, heaters, etc., and moving parts.
    c. Any uninsulated current-carrying metal parts are secured and spaced adequately from all non-current carrying metal parts.
    d. All electrical ground, both external and internal to the product are correctly and securely connected.
    e. All water connections are properly tightened.
    f. All panels and covers are properly and securely reassembled.

5. Do not attempt an appliance repair if you have any doubts as to your ability to complete it in a safe and satisfactory manner.

MASTER PUBLICATIONS

TABLE OF CONTENTS

SECTION 3: SERVICE PROCEDURE
Component COMPONENT DATA

INTRODUCTION

The primary object of the serviceman should be to make necessary repairs or replacements, to put the dishwasher into operation as soon as possible, and to give the customer uninterrupted, trouble-free and efficient service. It is with this in mind that this service manual is written.

We have included the general principles of operation and the construction and service procedures in making replacement of repair parts when necessary.

Read this manual carefully. It was prepared to help you.

Follow the instructions for installation, testing and repair . . . and your work will be made much faster and, in most cases, much easier.

The following diagnosis chart is intended to be only a starting point in proceeding with servicing undercounter or portable dishwashers. The diagnosis chart can deal only in generalities; to effectively service any dishwasher, the serviceman must thoroughly understand the mechanical functions and electrical circuitry of the appliance.

A considerable amount of time and money can be saved if a serviceman will allow himself the time to properly analyze the probable cause of malfunction, before proceeding to remove any parts. Always be sure, first, that the machine is properly installed and its power is properly fused. Be sure that the hot or hot and cold water is being supplied to the dishwasher and the operator understands the function of the various controls and selectors.

Always visually check for burnt or broken terminals or connections before using electrical test equipment. Before attempting removal of any electrical part from the dishwasher, disconnect the power source. If a voltmeter or test lamp is being used for testing, the power source must be connected and extreme caution should be observed.

WASHABILITY FACTORS AND COMPLAINTS

It has been customary in the past that no matter what the washability complaint is, we tend to look for deficiencies in the mechanical function of a dishwasher. But there are other factors that are important for sparkling clean dishes. A dishwasher may be mechanically perfect and perform well throughout every cycle of operation, but if the water is not hot or the water is excessively hard, or the detergent you use is old and hard and does not readily dissolve, then you have washability problems.

Here are a few hints that may help:

1. Before turning on your dishwasher, run the hot water tap until the water gets hot. If your heater is located far from your dishwasher, the water will be luke warm on the initial fill period.

2. Most people would not have a water hardness kit to check how many G.P.G. (grains per gallon) in your water. Your local water softner company will be glad to test it for you.

3. Place some of the detergent you use in a glass of hot water. Does the detergent readily dissolve? Does it float like a big cake of soap or do you really have to shake or stir it to make it dissolve? If your detergent does not emulsify almost immediately, change your detergent. If your detergent does not readily rinse off of the glass, change your detergent.

4. Check each operation or cycle of your dishwasher.

(a) **The fill period.** When water stops running into the dishwasher after the fill period, open the door. The water level should be approximately ½ inch above the heating element.

(b) **The wash cycle.** During the wash cycle open the door very slowly. The dishwasher will stop. Water should not be overflowing the tub. Examine the dishes in all areas. Are they being saturated or is food still clinging to them? Too much water or too little water will cause a washability complaint. Stacking the dishes the wrong way will prevent other dishes from being washed properly.

(c) **The drain cycle.** Is water being left in the tub after drain? Check your food particles screen. Is it clean?

(d) **The detergent cups.** Are they completely empty after the wash cycle, or are they caked up? Does the water get to the detergent cups to wash them out?

(e) **The wetting agent.** Does the water spread over your dishes or does it leave spots? Perhaps the wetting agent you use, if it is not the detergent, is at fault.

IF YOU FOLLOW THE ABOVE, YOUR WASHABILITY PROBLEMS WILL DISAPPEAR.

TROUBLE DIAGNOSIS CHART

PROBLEM	POSSIBLE CAUSE	REMEDY
1. Machine will not run at all.	1a. No power to machine. b. Loose leads. c. Door switch not operating. d. Timer not operating. e. Timer switch not energized.	1a. Check for power to outlet. b. Check and secure all leads. c. Check door adjustment and continuity through switch. d. Check out timer. e. Try switch for action.
2. Does not make a complete cycle.	2a. No power to inoperative components. b. Timer operating erratically. c. Loose lead. d. Defective components. e. Thermostat defective or not making physical contact with tub.	2a. Check for power leading to component. b. Check operation against sequence chart on wiring diagram. c. Check and secure all leads. d. Check for proper operation. e. Check thermostat for proper operation.
3. Water does not enter machine.	3a. Water supply valves closed. b. Open circuit in wiring. c. Timer not operating. d. Solenoid not operating. e. Supply line restricted. f. Pressure switch faulty.	3a. Open valve. b. Check continuity. c. Check out timer. d. Check leads and operation. e. Check for kinks and foreign material in lines. f. Check out.
4. Water does not drain from machine (within normal drain period).	4a. Restricted lines. b. Pump jammed. c. Motor not reversing.	4a. Check for kinks and foreign matter. (If connected to disposal check at drain port) b. Remove foreign material. c. Check timer switching and condition of the sub-interval switch (pump should run counterclockwise during pump out).
5. Water leakage.	5a. Poor door seal. b. Fill tube over spray or splash. c. Split hose or loose clamps. d. Pump leakage. e. Overfill. f. Tub leaks. g. Heater fasteners.	5a. Adjust latch. Check gasket. b. Check alignment of tube and end of tube for burrs or deposits left by hard water. c. Check condition of hoses and clamps. d. Replace pump assembly. e. Check operation of pressure switch, timer and inlet valve. f. Repair with Epoxi Patch Kit. g. Tighten nuts, replace washers.

PROBLEM	POSSIBLE CAUSE	REMEDY
5. Water leakage. (cont'd.)	5h. Pump not properly sealed.	5h. Reposition pump on gasket, tighten screws.
	i. Pressure switch.	i. Tighten nuts evenly around switch.
6. Poor washability	6a. Improper water level. (Proper level should submerge the heater.)	6a. Check level when fill cycle is complete. Check flow washer, water pressure and installation of drain line (see checking procedures.)
	b. Improper water temperature.	b. Check temperature in machine during last rinse (140° to 160°).
	c. Undesirable water conditions. (Hardness — excess iron).	c. Recommend water softener.
	d. Improper use of detergent.	d. Recommend change of brands and varying the amounts used.
	e. Improper loading of dishes.	e. Instruct customer on proper procedure.
	f. Detective components.	f. See below:
	1. Pump motor not operating.	1. Check for restrictions; electrical circuit — replace if necessary.
	2. Spray arm not turning.	2. Check for binding. Check clearance between arm and basket, rails, etc.
	3. Detergent dispenser not dumping.	3. Check coils and electrical circuit.
	g. Washing with dirty water.	g. See below:
	1. Loose or dirty filter screen.	1. Refit filter screen to eliminate gaps; wash out thoroughly.
	2. Drain pump inefficient.	2. Check to see if water is being evacuated. Look for obstruction, kinks, foreign material and motor operation.
	3. Back siphoning from sink.	3. Check installation, drain loop and/or drain air gap.
	h. Incorrect timer function.	h. Check timer against sequence chart.
	i. Incorrect lower spray arm and/or hub (two level spray arm models only)	i. Check lower spray arm for a Figure "3" stamped on the top surface. Check interior of hub for white colored flow straightener.
7. Abnormal noise (other than that caused by water striking dishes and tub sides).	7a. Spray arm hitting.	7a. Check clearances and loading habit of customer. Check hub and nut to be sure they are tight.
	b. Foreign matter in tub.	b. Check for broken dishes, etc.

PROBLEM	POSSIBLE CAUSE	REMEDY
7. Abnormal noise (cont'd.)	7c. Foreign matter in pump. d. Loose parts. e. Low water level. f. Improper loading of machine. g. Machine not level and solid.	7c. Remove from pump. d. Check for loose parts. e. Check for proper level. (Should submerge the heater.) f. Instruct customer. g. Check installation (see leveling instructions).
8. Unsatisfactory drying.	8a. Low water temperature. b. Heater element inoperative. c. Improper use by customer. d. Incomplete drain out.	8a. Should be 140° in machine at last rinse. b. Check electrical circuit, replace heater if defective. c. Explain drying method to customer. d. Refer to Item 4.
9. Overfill of water.	9a. Incorrect flow washer. b. Malfunction of fill valve. c. Incomplete drain out. d. Pressure switch not functioning properly.	9a. Check and install correct washer. b. Check: continuity to solenoid, timer function and foreign material holding plunger open. c. Refer to Item 4. d. Check switch.
10. Overload keeps tripping out.	10a. Open start winding. b. Defective relay. c. Tight motor bearing. d. Locked rotor. e. Grounded winding. f. Low voltage.	10a. Check continuity at the motor. b. Install known good relay. c. Check rotation of motor manually. (Turn off power first.) d. Check for obstructions. e. Check continuity of motor. f. Check voltage in house.
11. Underfill of water. (Proper water level should submerge the heater) - Some models use "low fill" gentle wash. (See "Specifications").	11a. Insufficient water supply. b. Defective or wrong flow washer. c. Restricted valve screen. d. Partially restricted feed line. e. Low fill gentle wash inoperative when "Gentle Wash" or "China Crystal" cycle is selected. Fill level will be ½" below heater. f. Pressure switch.	11a. Check water pressure and volume being delivered. b. Check flow washer. c. Check filter screen for foreign material. d. Check for kinks and foreign matter. e. Check unit in "Super Wash" for normal fill level. (Proper water level should submerge the heater). f. Check calibration.
12. Damaged Porcelain.	12a. Customer abuse. b. Shipping damage.	12a. Patch with Epoxi Kit — No. 674890 for stippled surfaces.

CYCLE CHART

NOTES.
A ONLY 1 MIN. FILL IN "CHINA CRYSTAL"
B WET. AGT. DISPENSES WHEN BOTH HEATER & DET. DISP. RELAYS ARE ENERGIZED.
C POTS & PANS ONLY.
D "HIGH TEMP. WASHING" ONLY.
E HEATER OFF WHEN "AIR DRY" IS SELECTED. HEATER CYCLES ON & OFF IN "CHINA CRYSTAL".
F DRAIN & FILL TO PURGE WATER LINES IN "QUICK WASH" ONLY.
WASH
RINSE
RINSE
WASH
RINSE
RINSE
DRY (HEAT OPTIONAL)
DRY-NO HEAT
T. HOLD POTS & PANS ONLY
DRAIN
DRAIN
DRAIN
DET. DISP.
DRAIN
DRAIN
WET. AGT. DISP./T. HOLD
DRAIN
FILL
FILL
FILL
FILL
HEAT ON
FILL
FILL
DIGITAL DISPLAY
CYCLES
POTS & PANS
NORMAL HEAVY
NORMAL LIGHT
CHINA CRYSTAL
DIGITAL DISPLAY QUICK WASH
DIGITAL DISPLAY RINSE HOLD
(HEATER CYCLES ON AND OFF)
(RINSE L.E.D. ON IN THIS CYCLE)
L.E.D'S.
WASH L.E.D.
RINSE L.E.D.
DRY L.E.D.
HOLD L.E.D.
RELAY OUTPUTS
DET. DISP.
MAIN
MOTOR
WASH/DN.
FILL
HEATER
(X = OUTPUT OFF 4 SEC., ON 56 SEC.)
POWER SUPPLY EDGE CONN.
Y-R CLN ON
Y-BK CLN OFF
NOT USED
NOT USED
BR-W V RELAY
OR TR 2
OR-W TR 2
G GROUND
LATCH
BU-BK CLOCK
R TR 1
R-W TR 1
NOT USED
W-BK V SS
R-BK V GG
DISPLAY EDGE CONNECTOR
R-OR
LATCH
W
MICROCOMPUTER EDGE CONNECTOR
BU-BK CLK
R-BK Vgg
Y-BK CLN OFF
NOT USED
Y-W MAIN
W-G MOTOR
BU-Y WASH/DN
O-BU DET. DISP.
LATCH
BU-GY HEATER
R-BU FILL VALVE
Y-R CLN ON
BR-O RELAY COM
BR-W V RELAY
NOT USED
W-BK V SS
NOT USED
O-GY THERMOSTAT
O-T DOOR SW.
W-BU DOOR SW.
T-R THERMOSTAT
P/S-DISPLAY EDGE CONN.
R-OR
LATCH
W
NOT USED
BK
TO TOUCH MODULE
1
2
3
4
5
6
7
8
FLEX CIRCUIT CONNECTOR

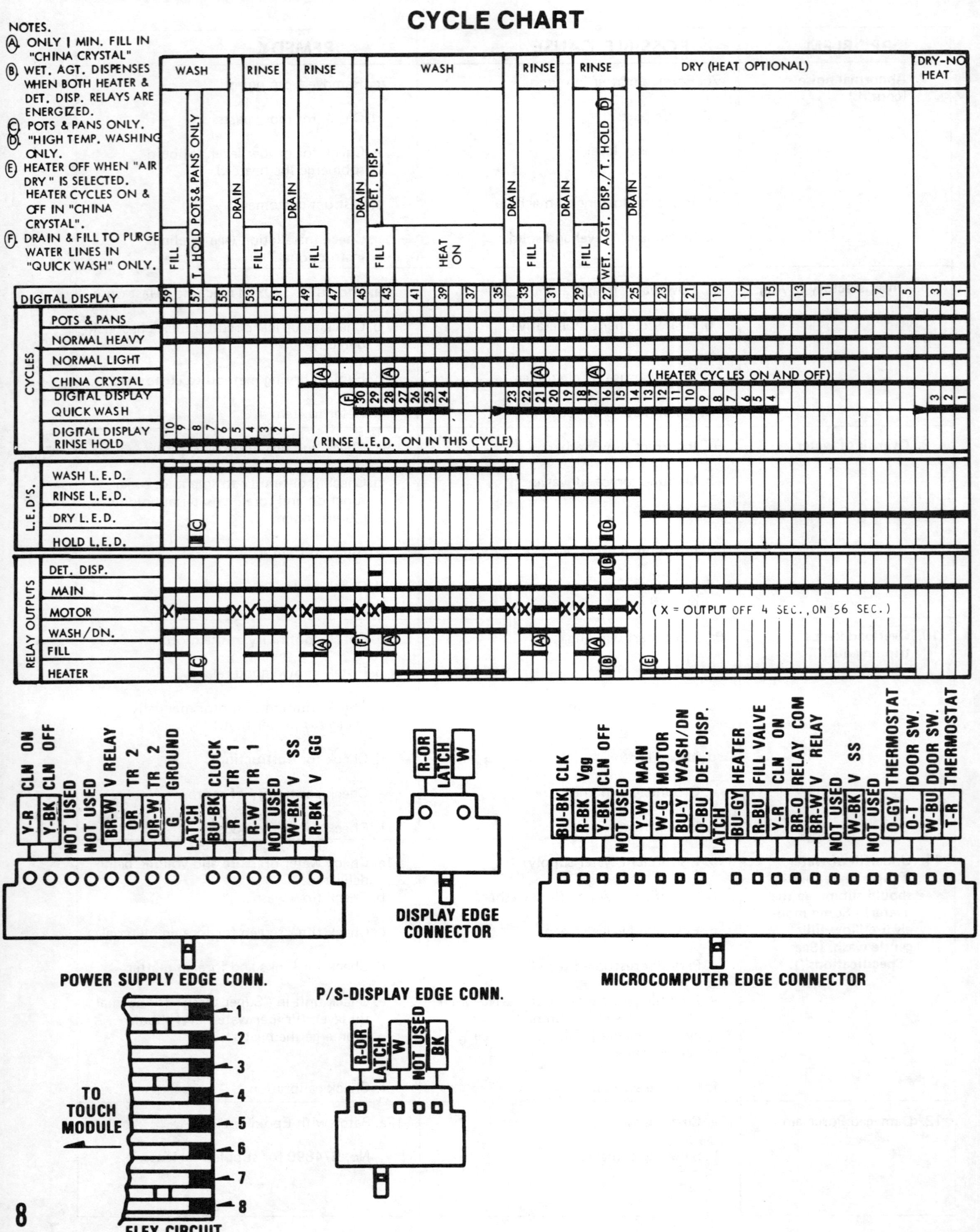

TROUBLE DIAGNOSIS CHART

ELECTRONIC CONTROL MODELS, tests

CAUTION: Disconnect power supply at the fuse box before attempting to replace a component.

There are three test cycles built into the microcomputor which energizes all major components for use in diagnosing problems.

Test Cycle #1, refer to Figure 1

1. Press and hold "Quick Wash".
2. Press and hold "Pots and Pans".
3. While holding 1 and 2 above, press and hold "Start" until "03" is displayed. Test can be cancelled by touching "Drain/Off". Touch once to pump water out, and twice for no pump out.

1. Every second, the digital display will change numbers "00", "11", "22", "33", "44", "55", "66", "77", "88", "99".
2. From "55" thru "99" the five display L.E.D.s will light, these are Wash, Rinse, Dry, Hold and Clean.
3. The pattern continually repeats until the cycle is cancelled.

Test Cycle #3

1. Press and hold "Quick Wash".
2. Press and hold "Normal Light".
3. While holding 1 and 2 above, press and hold "Start" until "dd" is displayed.

DIGITAL DISPLAY AND L.E.D.s — RELAYS ENERGIZED

DIGITAL DISPLAY AND L.E.D.s	RELAYS ENERGIZED
"03" Digital Display "Wash" L.E.D. on (1 minute)	Main Fill Motor Run Wash/Drain Detergent Dispenser* Heater* *Both on at the same time gives rinse aid dispenser.
"02" Digital Display "Wash" and "Hold" L.E.D.s on	Main, Motor Run Wash/Drain
Pressing "China Crystal" and "Rinse Hold" at the same time will terminate washing cycle. Motor stops and Wash/Drain relay is de-energized. The motor will then restart in drain.	
"02" Digital Display "Rinse" L.E.D. on (1 minute)	Main Motor Run
"01" Digital Display "Dry" L.E.D. on (30 seconds)	Main Heater
Digital Display off "Clean" L.E.D. on	Close and Open tilt door to turn "Clean" L.E.D. off.

Figure 1, test charts

Test Cycle #2

1. Press and hold "Quick Wash".
2. Press and Hold "Normal Heavy".
3. While holding 1 and 2 above, press and hold "Start" until "11" is displayed.

Test can be cancelled by pressing "Drain/Off" twice.

This cycle energizes the digits and L.E.D.s on the digital display assembly as follows:

"dd" digital display and "Hold" L.E.D. will remain on until "China Crystal" and "Rinse Hold" are pressed and held at the same time that "Clean" L.E.D. is observed. To turn the "Clean" L.E.D. off, close and open the tilt door. This cycle simulates the thermostat input for "Pots and Pans" and "Hi—Temp Washing" water heating periods.

NOTE: After the diagnosis and the reparing of the unit, run through Test Cycles #1 and 2 to check performance

ELECTRONIC CONTROL MODELS

Touch Module Continuity Check, referenced to continuity diagram.

1. Press selection desired to be checked.

2. Check continuity between the correct bands on the flex connector. EXAMPLE: Normal Heavy — check bands 1 and 4 while pressing the selection pad.

DIAGNOSIS PROCEDURE

All symptoms of a problem should be carefully noted. A users description of a problem may not be complete. therefore it is recommended that the following short tests be completed before looking for a specific problem.

1. Open tilt door. Observe which L.E.D.s and digital display numbers. L.E.D.s should be "Normal Heavy" and "Heat Dry". Digital Display should be "00".

2. Press each of the six cycle selection pads and watch for the L.E.D. in each pad to light.

3. Open and close the tilt door. Touch the "Hy-Temp. Washing" and "Heat Dry" pads several times. The indicator L.E.D.s should turn on and off with repeated touches.

4. Touch "Delay Start" and watch for its L.E.D. to light.

5. Touch "Drain/Off" twice within four seconds. First touch should light its L.E.D. The second touch should turn off the L.E.D. and avoid a pump out.

6. Press "Start". If the fill valve energizes, press "Drain/Off" twice.

7. If the user has a specific complaint regarding failure of a component, use the test cycles above to verify the complaint.

8. Check all of the connectors, terminals, wiring harnesses for obvious breaks and burnt spots Look for loose terminals that may cause an intermittent problem.

TOUCH MODULE CONTINUITY DIAGRAM

	1	3	5	7
2	POTS & PANS	HI-TEMP WASHING	NOT USED	NOT USED
4	NORMAL HEAVY	HEAT DRY	DRAIN/OFF	QUICK WASH
6	NOT USED	NOT USED	START DELAY WASH	CHINA CRYSTAL
8	NORMAL LIGHT	NOT USED	START	RINSE HOLD

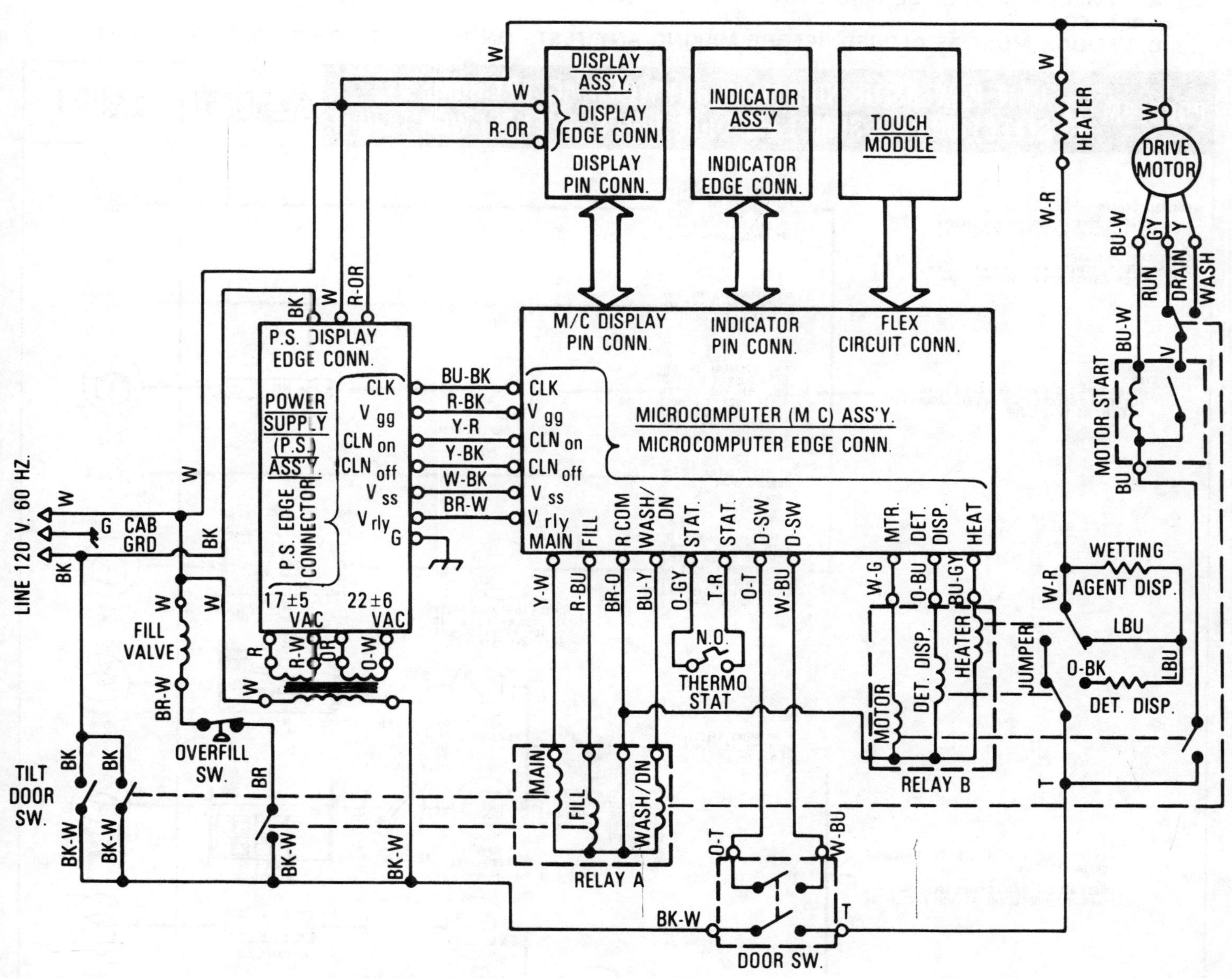
DISPLAY ASS'Y.
DISPLAY EDGE CONN.
DISPLAY PIN CONN.
INDICATOR ASS'Y
INDICATOR EDGE CONN.
TOUCH MODULE
HEATER
DRIVE MOTOR
W
R-OR
W-R
BU-W
RUN
GY
DRAIN
Y
WASH
MOTOR START
BU-W
V
BU
M/C DISPLAY PIN CONN.
INDICATOR PIN CONN.
FLEX CIRCUIT CONN.
BK
W
R-OR
P.S. DISPLAY EDGE CONN.
POWER SUPPLY (P.S.) ASS'Y.
P.S. EDGE CONNECTOR
CLK
V gg
CLN on
CLN off
V ss
V rly
G
BU-BK
R-BK
Y-R
Y-BK
W-BK
BR-W
CLK
V gg
CLN on
CLN off
V ss
V rly
MICROCOMPUTER (M C) ASS'Y.
MICROCOMPUTER EDGE CONN.
MAIN
FILL
R COM
WASH/DN
STAT.
STAT.
D-SW
D-SW
MTR.
DET. DISP.
HEAT
17±5 VAC
22±6 VAC
R-W
OR
O-W
LINE 120 V. 60 HZ.
W
G CAB GRD
BK
BK
W
W
FILL VALVE
BR-W
W
R
W
BR
OVERFILL SW.
TILT DOOR SW.
BK
BK
BK
BK-W
BK-W
BK-W
BK-W
BK-W
Y-W
R-BU
BR-O
BU-Y
O-GY
T-R
O-T
W-BU
N.O.
THERMO STAT
MAIN
FILL
WASH/DN
RELAY A
O-T
W-BU
BK-W
T
DOOR SW.
W-G
O-BU
BU-GY
MOTOR
DET. DISP.
HEATER
RELAY B
W-R
JUMPER
O-BK
WETTING AGENT DISP.
LBU
LBU
DET. DISP.
T

SECTION 1.

DIAGNOSIS PROCEDURE

1. A DIAGNOSIS OF THIS D/W MUST BEGIN WITH NORMAL CHECKS OF LINE VOLTAGE, BLOWN FUSES AND DEFECTIVE COMPONENTS.
2. ALL CHECKS SHOULD BE MADE WITH A METER HAVING A SENSITIVITY OF 20,000 OHMS PER VOLT OR GREATER.
3. D/W DOOR MUST BE CLOSED BEFORE MAKING ANY TESTS ON THE ELECTRONIC CONTROL SYSTEM.

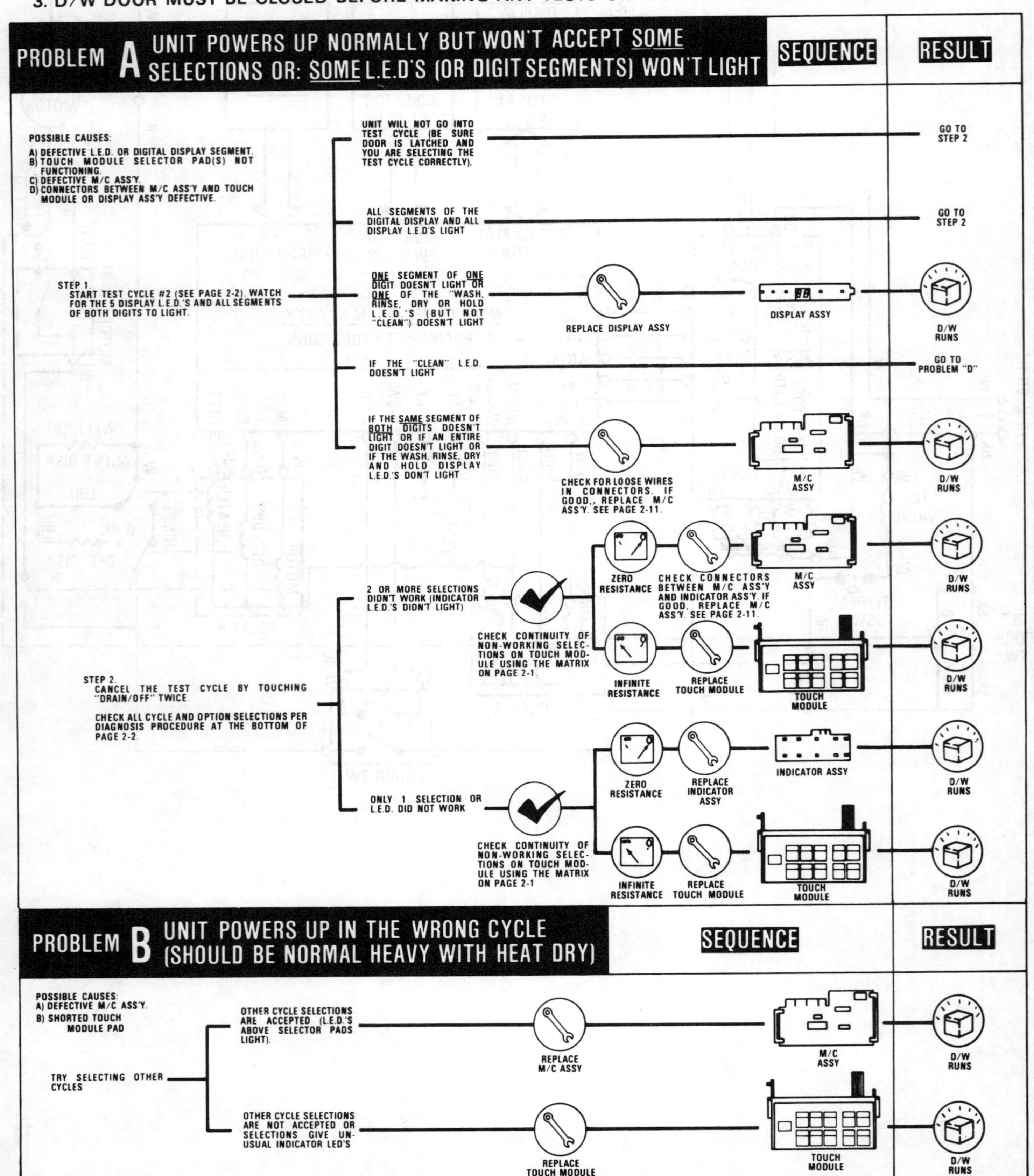

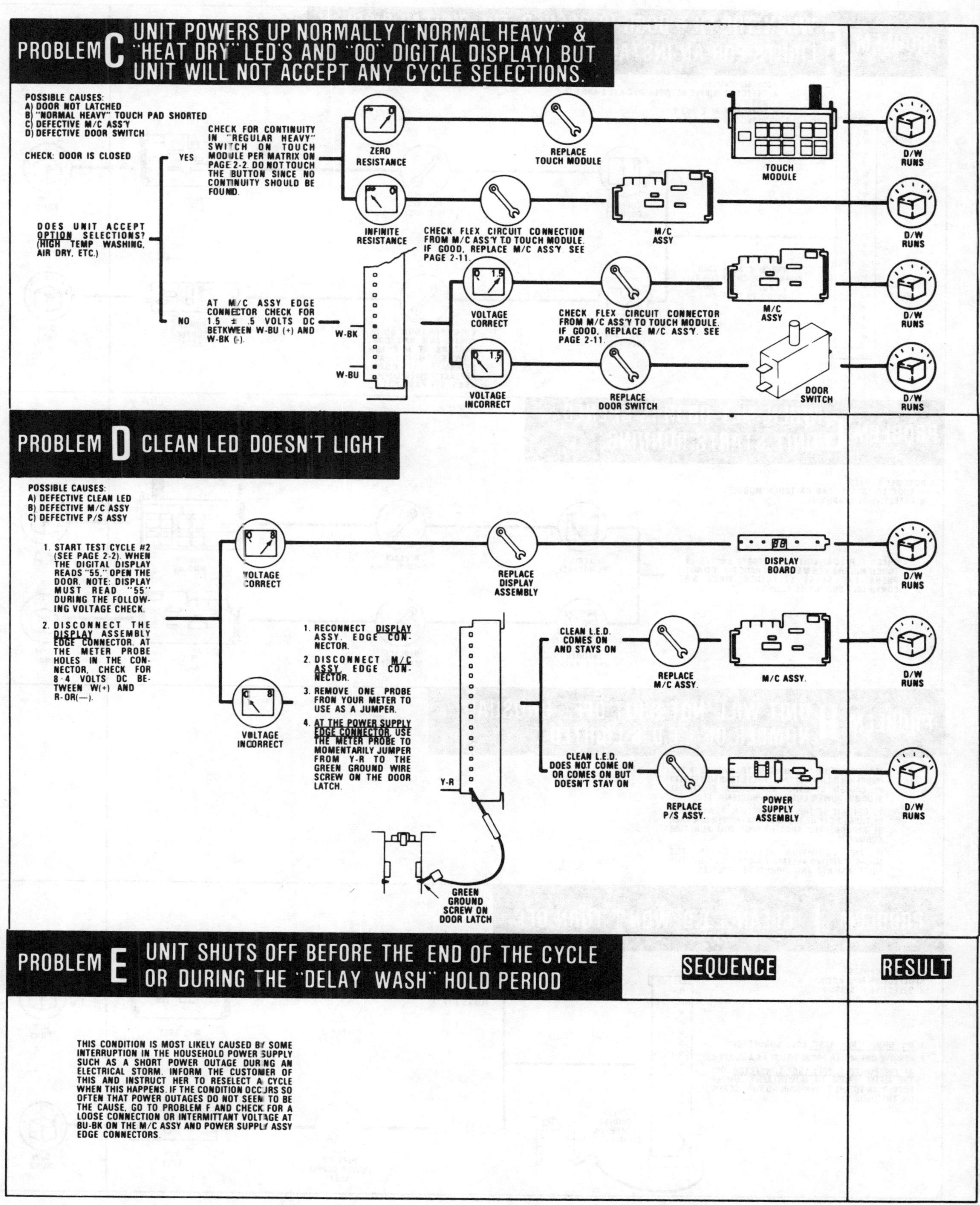

PROBLEM C UNIT POWERS UP NORMALLY ("NORMAL HEAVY" & "HEAT DRY" LED'S AND "00" DIGITAL DISPLAY) BUT UNIT WILL NOT ACCEPT ANY CYCLE SELECTIONS.

POSSIBLE CAUSES:
A) DOOR NOT LATCHED
B) "NORMAL HEAVY" TOUCH PAD SHORTED
C) DEFECTIVE M/C ASS'Y
D) DEFECTIVE DOOR SWITCH

CHECK: DOOR IS CLOSED

DOES UNIT ACCEPT OPTION SELECTIONS? (HIGH TEMP WASHING, AIR DRY, ETC.)

YES

NO

CHECK FOR CONTINUITY IN "REGULAR HEAVY" SWITCH ON TOUCH MODULE PER MATRIX ON PAGE 2-2. DO NOT TOUCH THE BUTTON SINCE NO CONTINUITY SHOULD BE FOUND.

ZERO RESISTANCE

INFINITE RESISTANCE

REPLACE TOUCH MODULE

CHECK FLEX CIRCUIT CONNECTION FROM M/C ASS'Y TO TOUCH MODULE. IF GOOD, REPLACE M/C ASS'Y SEE PAGE 2-11.

TOUCH MODULE

M/C ASS'Y

D/W RUNS

AT M/C ASSY EDGE CONNECTOR CHECK FOR 1.5 ± .5 VOLTS DC BETWEEN W-BU (+) AND W-BK (-).

W-BK

W-BU

VOLTAGE CORRECT

VOLTAGE INCORRECT

CHECK FLEX CIRCUIT CONNECTOR FROM M/C ASS'Y TO TOUCH MODULE. IF GOOD, REPLACE M/C ASS'Y. SEE PAGE 2-11.

REPLACE DOOR SWITCH

M/C ASS'Y

DOOR SWITCH

D/W RUNS

PROBLEM D CLEAN LED DOESN'T LIGHT

POSSIBLE CAUSES:
A) DEFECTIVE CLEAN LED
B) DEFECTIVE M/C ASSY
C) DEFECTIVE P/S ASSY

1. START TEST CYCLE #2 (SEE PAGE 2-2). WHEN THE DIGITAL DISPLAY READS "55," OPEN THE DOOR. NOTE: DISPLAY MUST READ "55" DURING THE FOLLOWING VOLTAGE CHECK.

2. DISCONNECT THE DISPLAY ASSEMBLY EDGE CONNECTOR. AT THE METER PROBE HOLES IN THE CONNECTOR, CHECK FOR 8.4 VOLTS DC BETWEEN W(+) AND R-OR(—).

VOLTAGE CORRECT

VOLTAGE INCORRECT

REPLACE DISPLAY ASSEMBLY

DISPLAY BOARD

D/W RUNS

1. RECONNECT DISPLAY ASSY. EDGE CONNECTOR.
2. DISCONNECT M/C ASSY EDGE CONNECTOR.
3. REMOVE ONE PROBE FROM YOUR METER TO USE AS A JUMPER.
4. AT THE POWER SUPPLY EDGE CONNECTOR, USE THE METER PROBE TO MOMENTARILY JUMPER FROM Y-R TO THE GREEN GROUND WIRE SCREW ON THE DOOR LATCH.

Y-R

GREEN GROUND SCREW ON DOOR LATCH

CLEAN L.E.D. COMES ON AND STAYS ON

CLEAN L.E.D. DOES NOT COME ON OR COMES ON BUT DOESN'T STAY ON

REPLACE M/C ASSY.

REPLACE P/S ASSY.

M/C ASSY.

POWER SUPPLY ASSEMBLY

D/W RUNS

PROBLEM E UNIT SHUTS OFF BEFORE THE END OF THE CYCLE OR DURING THE "DELAY WASH" HOLD PERIOD

SEQUENCE

RESULT

THIS CONDITION IS MOST LIKELY CAUSED BY SOME INTERRUPTION IN THE HOUSEHOLD POWER SUPPLY SUCH AS A SHORT POWER OUTAGE DURING AN ELECTRICAL STORM. INFORM THE CUSTOMER OF THIS AND INSTRUCT HER TO RESELECT A CYCLE WHEN THIS HAPPENS. IF THE CONDITION OCCURS SO OFTEN THAT POWER OUTAGES DO NOT SEEM TO BE THE CAUSE, GO TO PROBLEM F AND CHECK FOR A LOOSE CONNECTION OR INTERMITTANT VOLTAGE AT BU-BK ON THE M/C ASSY AND POWER SUPPLY ASSY EDGE CONNECTORS.

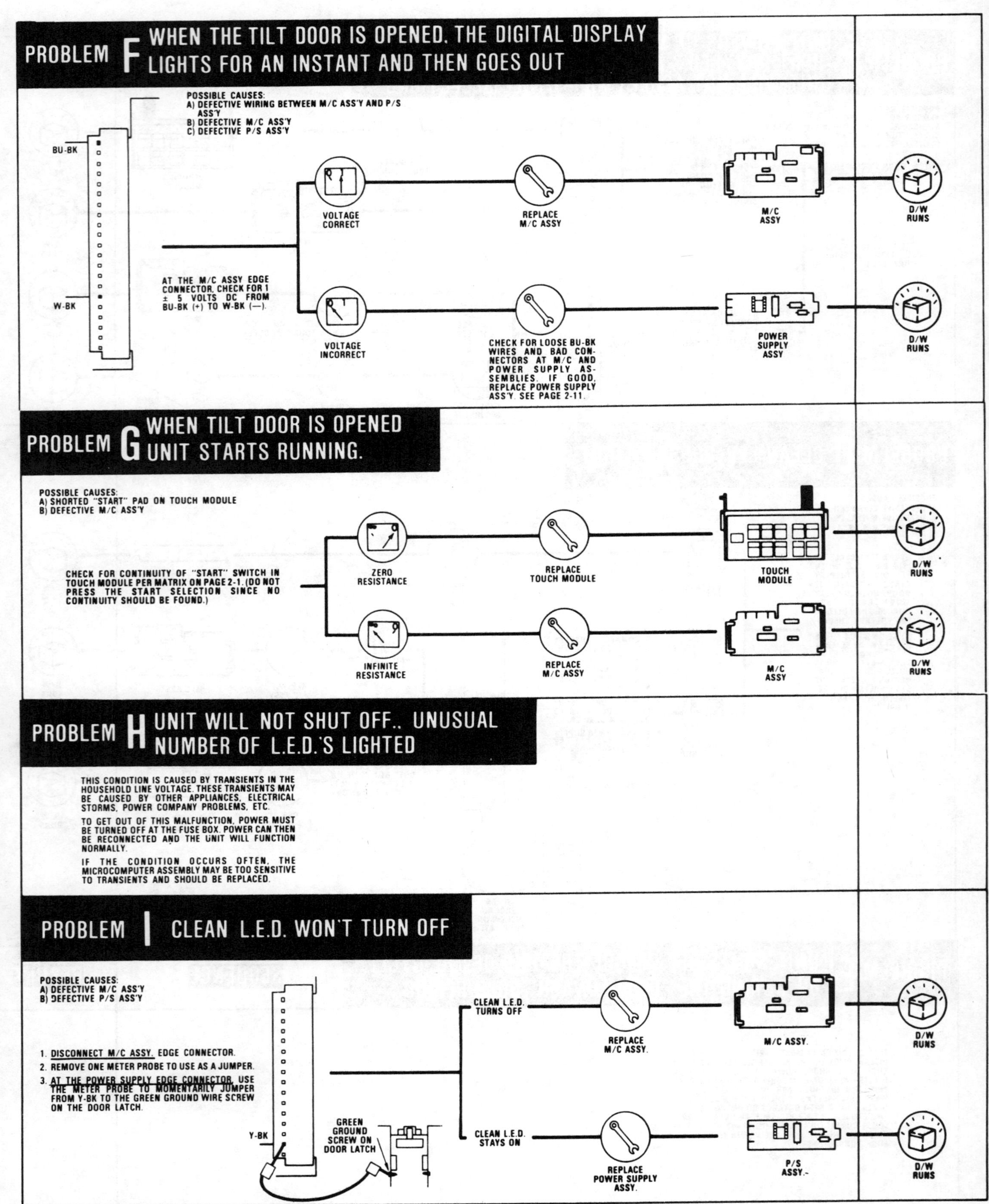
PROBLEM F WHEN THE TILT DOOR IS OPENED. THE DIGITAL DISPLAY LIGHTS FOR AN INSTANT AND THEN GOES OUT

POSSIBLE CAUSES:
A) DEFECTIVE WIRING BETWEEN M/C ASS'Y AND P/S ASS'Y
B) DEFECTIVE M/C ASS'Y
C) DEFECTIVE P/S ASS'Y

BU-BK
W-BK

AT THE M/C ASSY EDGE CONNECTOR, CHECK FOR 1 ± 5 VOLTS DC FROM BU-BK (+) TO W-BK (—).

VOLTAGE CORRECT
REPLACE M/C ASSY
M/C ASSY
D/W RUNS

VOLTAGE INCORRECT
CHECK FOR LOOSE BU-BK WIRES AND BAD CONNECTORS AT M/C AND POWER SUPPLY ASSEMBLIES. IF GOOD, REPLACE POWER SUPPLY ASS'Y. SEE PAGE 2-11.
POWER SUPPLY ASSY
D/W RUNS

PROBLEM G WHEN TILT DOOR IS OPENED UNIT STARTS RUNNING.

POSSIBLE CAUSES:
A) SHORTED "START" PAD ON TOUCH MODULE
B) DEFECTIVE M/C ASS'Y

CHECK FOR CONTINUITY OF "START" SWITCH IN TOUCH MODULE PER MATRIX ON PAGE 2-1. (DO NOT PRESS THE START SELECTION SINCE NO CONTINUITY SHOULD BE FOUND.)

ZERO RESISTANCE
REPLACE TOUCH MODULE
TOUCH MODULE
D/W RUNS

INFINITE RESISTANCE
REPLACE M/C ASSY
M/C ASSY
D/W RUNS

PROBLEM H UNIT WILL NOT SHUT OFF.. UNUSUAL NUMBER OF L.E.D.'S LIGHTED

THIS CONDITION IS CAUSED BY TRANSIENTS IN THE HOUSEHOLD LINE VOLTAGE. THESE TRANSIENTS MAY BE CAUSED BY OTHER APPLIANCES, ELECTRICAL STORMS, POWER COMPANY PROBLEMS, ETC.

TO GET OUT OF THIS MALFUNCTION, POWER MUST BE TURNED OFF AT THE FUSE BOX. POWER CAN THEN BE RECONNECTED AND THE UNIT WILL FUNCTION NORMALLY.

IF THE CONDITION OCCURS OFTEN, THE MICROCOMPUTER ASSEMBLY MAY BE TOO SENSITIVE TO TRANSIENTS AND SHOULD BE REPLACED.

PROBLEM I CLEAN L.E.D. WON'T TURN OFF

POSSIBLE CAUSES:
A) DEFECTIVE M/C ASS'Y
B) DEFECTIVE P/S ASS'Y

1. DISCONNECT M/C ASSY. EDGE CONNECTOR.
2. REMOVE ONE METER PROBE TO USE AS A JUMPER.
3. AT THE POWER SUPPLY EDGE CONNECTOR, USE THE METER PROBE TO MOMENTARILY JUMPER FROM Y-BK TO THE GREEN GROUND WIRE SCREW ON THE DOOR LATCH.

Y-BK
GREEN GROUND SCREW ON DOOR LATCH

CLEAN L.E.D. TURNS OFF
REPLACE M/C ASSY.
M/C ASSY.
D/W RUNS

CLEAN L.E.D. STAYS ON
REPLACE POWER SUPPLY ASSY.
P/S ASSY.-
D/W RUNS

PROBLEM J LED'S AND DIGITAL DISPLAY DO NOT LIGHT WHEN TILT DOOR IS OPENED

SEQUENCE

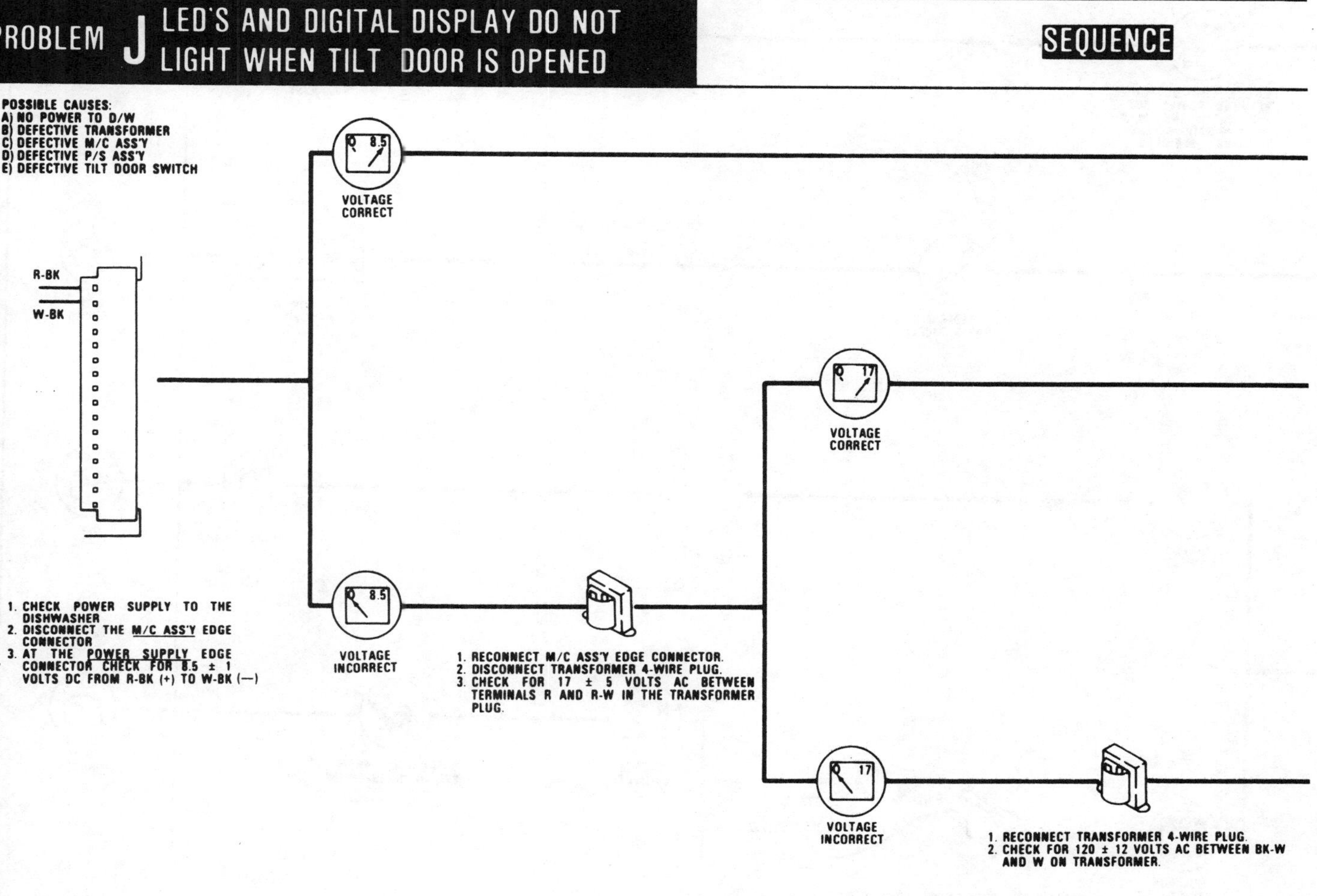

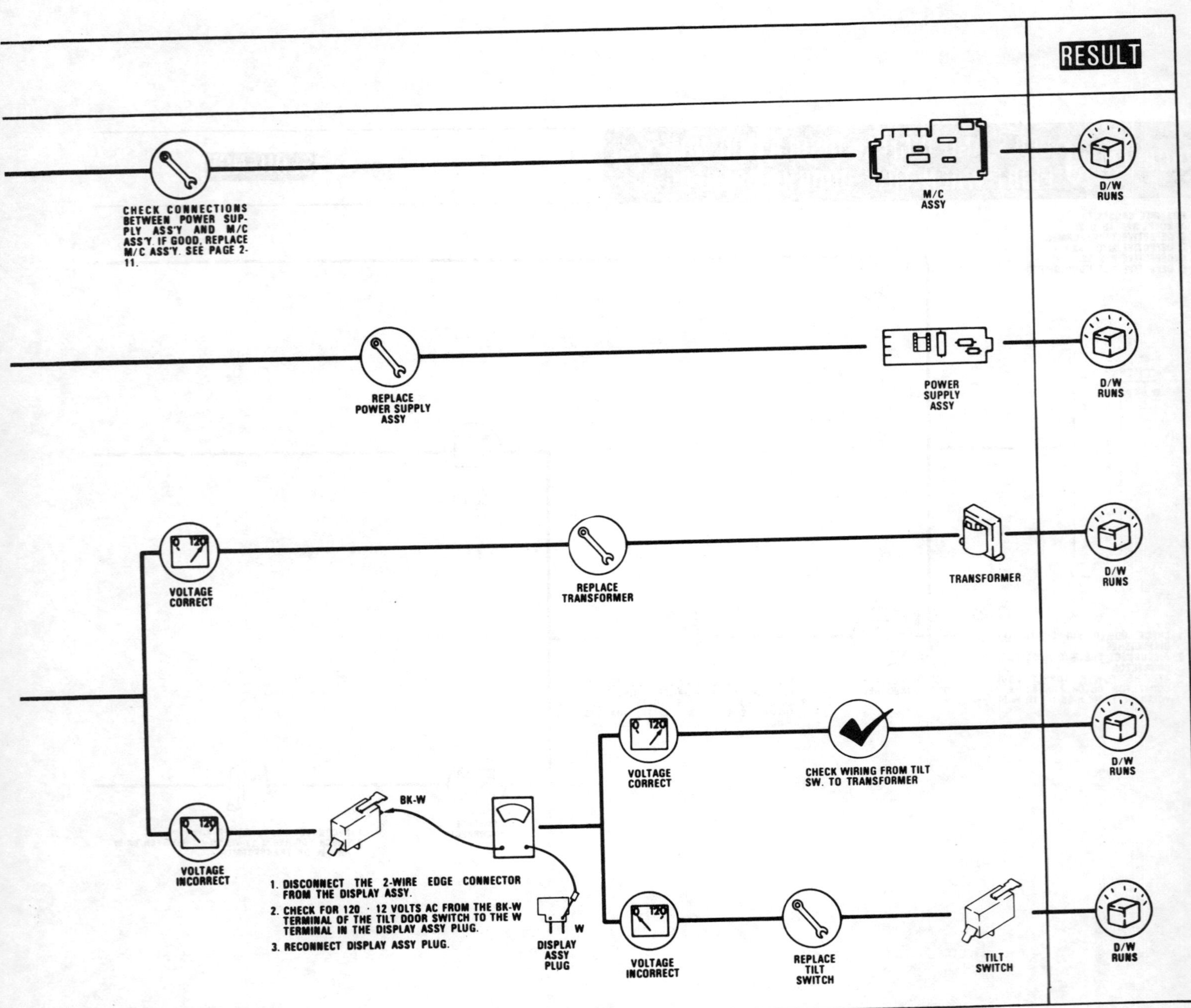
RESULT

CHECK CONNECTIONS BETWEEN POWER SUP-PLY ASS'Y AND M/C ASS'Y. IF GOOD, REPLACE M/C ASS'Y. SEE PAGE 2-11.
M/C ASSY
D/W RUNS

REPLACE POWER SUPPLY ASSY
POWER SUPPLY ASSY
D/W RUNS

VOLTAGE CORRECT
REPLACE TRANSFORMER
TRANSFORMER
D/W RUNS

VOLTAGE CORRECT
CHECK WIRING FROM TILT SW. TO TRANSFORMER
D/W RUNS

VOLTAGE INCORRECT
BK-W

1. DISCONNECT THE 2-WIRE EDGE CONNECTOR FROM THE DISPLAY ASSY.
2. CHECK FOR 120 - 12 VOLTS AC FROM THE BK-W TERMINAL OF THE TILT DOOR SWITCH TO THE W TERMINAL IN THE DISPLAY ASSY PLUG.
3. RECONNECT DISPLAY ASSY PLUG.
W
DISPLAY ASSY PLUG

VOLTAGE INCORRECT
REPLACE TILT SWITCH
TILT SWITCH
D/W RUNS

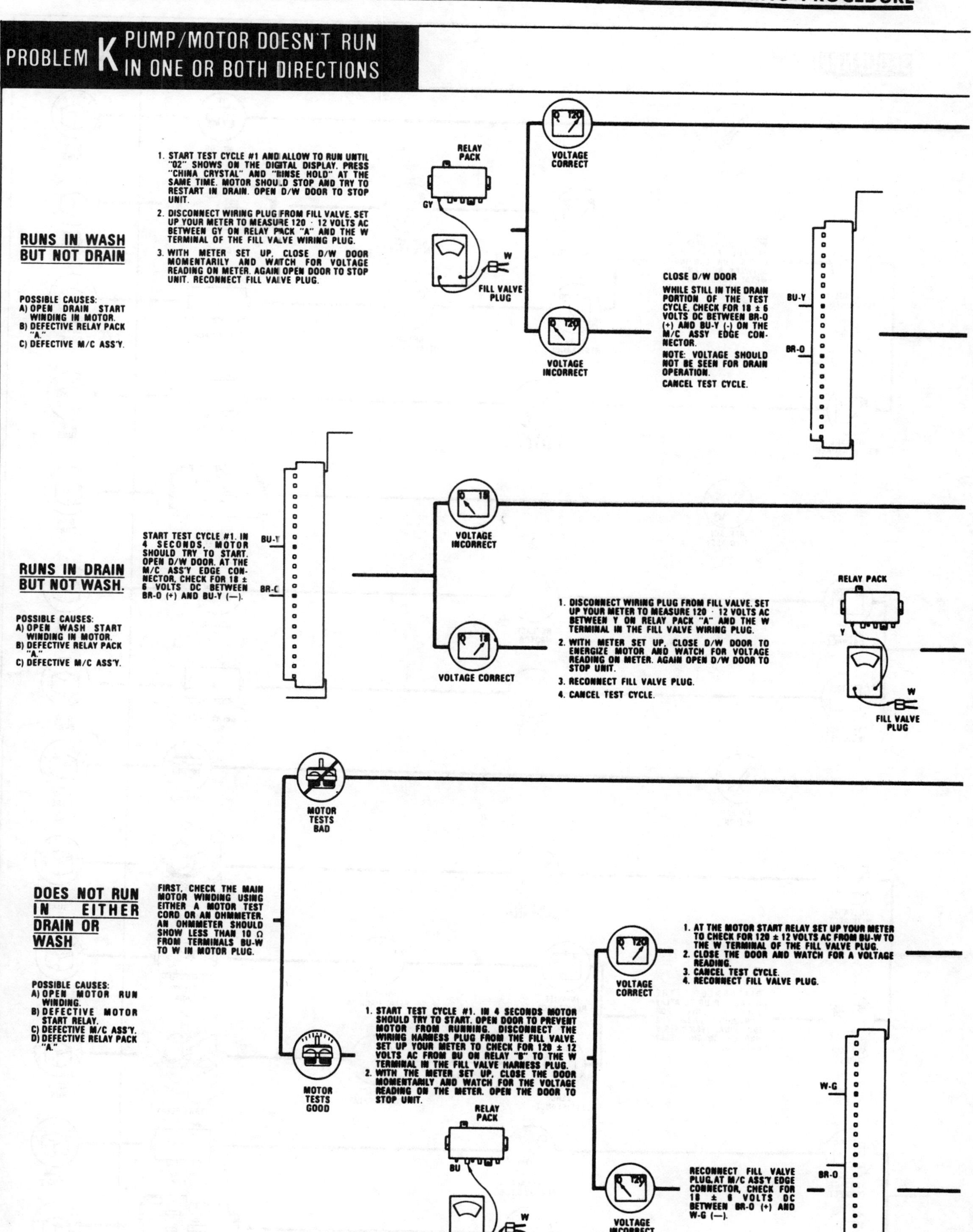
PROBLEM K PUMP/MOTOR DOESN'T RUN IN ONE OR BOTH DIRECTIONS

RUNS IN WASH BUT NOT DRAIN

POSSIBLE CAUSES:
A) OPEN DRAIN START WINDING IN MOTOR.
B) DEFECTIVE RELAY PACK "A."
C) DEFECTIVE M/C ASS'Y.

1. START TEST CYCLE #1 AND ALLOW TO RUN UNTIL "02" SHOWS ON THE DIGITAL DISPLAY. PRESS "CHINA CRYSTAL" AND "RINSE HOLD" AT THE SAME TIME. MOTOR SHOULD STOP AND TRY TO RESTART IN DRAIN. OPEN D/W DOOR TO STOP UNIT.
2. DISCONNECT WIRING PLUG FROM FILL VALVE. SET UP YOUR METER TO MEASURE 120 - 12 VOLTS AC BETWEEN GY ON RELAY PACK "A" AND THE W TERMINAL OF THE FILL VALVE WIRING PLUG.
3. WITH METER SET UP, CLOSE D/W DOOR MOMENTARILY AND WATCH FOR VOLTAGE READING ON METER. AGAIN OPEN DOOR TO STOP UNIT. RECONNECT FILL VALVE PLUG.

RELAY PACK
GY
W
FILL VALVE PLUG
VOLTAGE CORRECT
VOLTAGE INCORRECT

CLOSE D/W DOOR WHILE STILL IN THE DRAIN PORTION OF THE TEST CYCLE, CHECK FOR 18 ± 6 VOLTS DC BETWEEN BR-O (+) AND BU-Y (-) ON THE M/C ASS'Y EDGE CONNECTOR.
NOTE: VOLTAGE SHOULD NOT BE SEEN FOR DRAIN OPERATION.
CANCEL TEST CYCLE.

BU-Y
BR-O

RUNS IN DRAIN BUT NOT WASH.

POSSIBLE CAUSES:
A) OPEN WASH START WINDING IN MOTOR.
B) DEFECTIVE RELAY PACK "A."
C) DEFECTIVE M/C ASS'Y.

START TEST CYCLE #1. IN 4 SECONDS, MOTOR SHOULD TRY TO START. OPEN D/W DOOR. AT THE M/C ASS'Y EDGE CONNECTOR, CHECK FOR 18 ± 6 VOLTS DC BETWEEN BR-O (+) AND BU-Y (—).

BU-Y
BR-O

VOLTAGE INCORRECT
VOLTAGE CORRECT

1. DISCONNECT WIRING PLUG FROM FILL VALVE. SET UP YOUR METER TO MEASURE 120 - 12 VOLTS AC BETWEEN Y ON RELAY PACK "A" AND THE W TERMINAL IN THE FILL VALVE WIRING PLUG.
2. WITH METER SET UP, CLOSE D/W DOOR TO ENERGIZE MOTOR AND WATCH FOR VOLTAGE READING ON METER. AGAIN OPEN D/W DOOR TO STOP UNIT.
3. RECONNECT FILL VALVE PLUG.
4. CANCEL TEST CYCLE.

RELAY PACK
Y
W
FILL VALVE PLUG

DOES NOT RUN IN EITHER DRAIN OR WASH

POSSIBLE CAUSES:
A) OPEN MOTOR RUN WINDING.
B) DEFECTIVE MOTOR START RELAY.
C) DEFECTIVE M/C ASS'Y.
D) DEFECTIVE RELAY PACK "A."

FIRST, CHECK THE MAIN MOTOR WINDING USING EITHER A MOTOR TEST CORD OR AN OHMMETER. AN OHMMETER SHOULD SHOW LESS THAN 10 Ω FROM TERMINALS BU-W TO W IN MOTOR PLUG.

MOTOR TESTS BAD

MOTOR TESTS GOOD

1. START TEST CYCLE #1. IN 4 SECONDS MOTOR SHOULD TRY TO START. OPEN DOOR TO PREVENT MOTOR FROM RUNNING. DISCONNECT THE WIRING HARNESS PLUG FROM THE FILL VALVE. SET UP YOUR METER TO CHECK FOR 120 ± 12 VOLTS AC FROM BU ON RELAY "B" TO THE W TERMINAL IN THE FILL VALVE HARNESS PLUG.
2. WITH THE METER SET UP, CLOSE THE DOOR MOMENTARILY AND WATCH FOR THE VOLTAGE READING ON THE METER. OPEN THE DOOR TO STOP UNIT.

1. AT THE MOTOR START RELAY SET UP YOUR METER TO CHECK FOR 120 ± 12 VOLTS AC FROM BU-W TO THE W TERMINAL OF THE FILL VALVE PLUG.
2. CLOSE THE DOOR AND WATCH FOR A VOLTAGE READING.
3. CANCEL TEST CYCLE.
4. RECONNECT FILL VALVE PLUG.

VOLTAGE CORRECT

W-G

RELAY PACK
BU
W
FILL VALVE PLUG

VOLTAGE INCORRECT

RECONNECT FILL VALVE PLUG. AT M/C ASS'Y EDGE CONNECTOR, CHECK FOR 18 ± 6 VOLTS DC BETWEEN BR-O (+) AND W-G (—).

BR-O

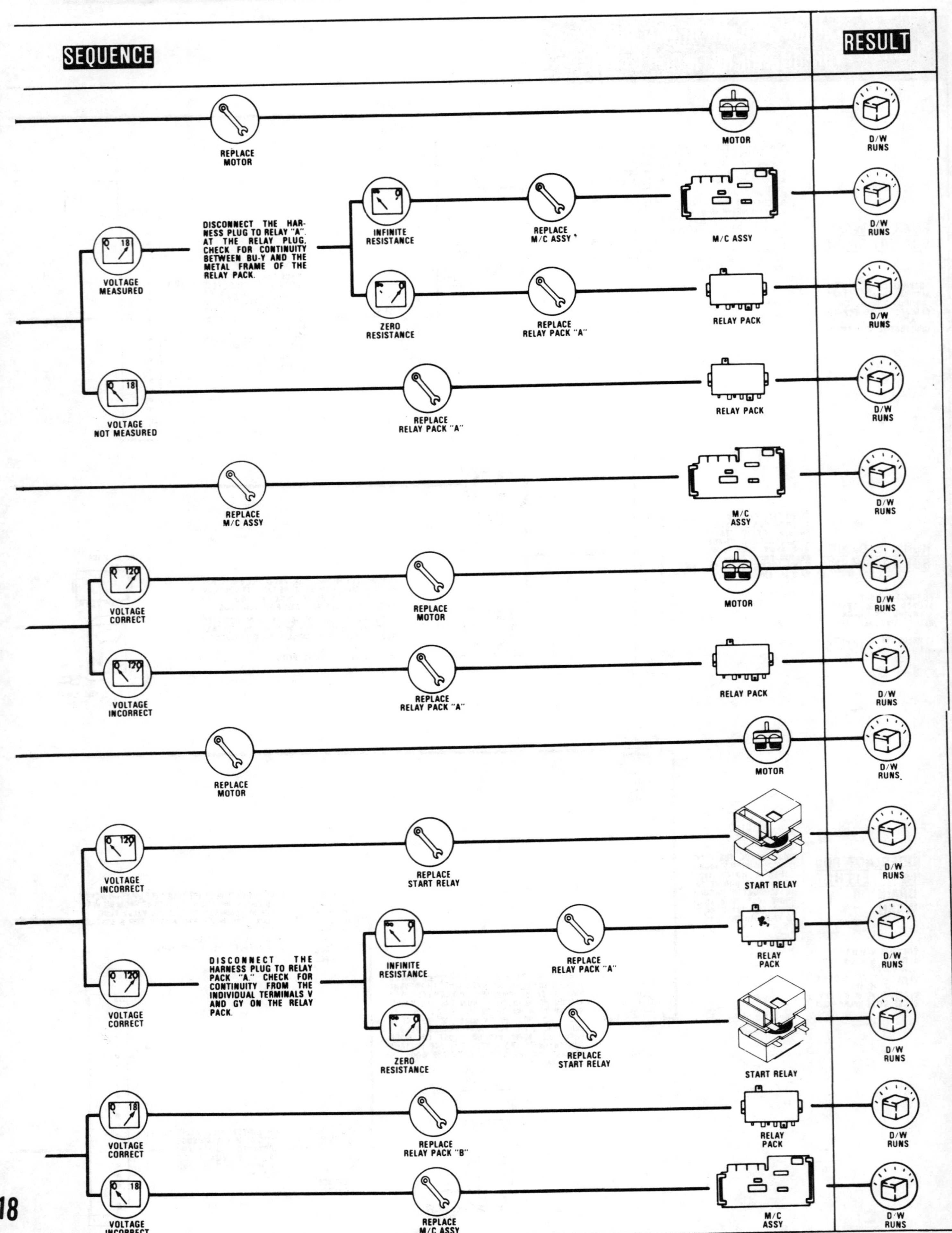
SEQUENCE

RESULT

REPLACE
MOTOR

MOTOR

D/W
RUNS

VOLTAGE
MEASURED

DISCONNECT THE HAR-
NESS PLUG TO RELAY "A".
AT THE RELAY PLUG,
CHECK FOR CONTINUITY
BETWEEN BU-Y AND THE
METAL FRAME OF THE
RELAY PACK.

INFINITE
RESISTANCE

REPLACE
M/C ASSY

M/C ASSY

D/W
RUNS

ZERO
RESISTANCE

REPLACE
RELAY PACK "A"

RELAY PACK

D/W
RUNS

VOLTAGE
NOT MEASURED

REPLACE
RELAY PACK "A"

RELAY PACK

D/W
RUNS

REPLACE
M/C ASSY

M/C
ASSY

D/W
RUNS

VOLTAGE
CORRECT

REPLACE
MOTOR

MOTOR

D/W
RUNS

VOLTAGE
INCORRECT

REPLACE
RELAY PACK "A"

RELAY PACK

D/W
RUNS

REPLACE
MOTOR

MOTOR

D/W
RUNS

VOLTAGE
INCORRECT

REPLACE
START RELAY

START RELAY

D/W
RUNS

DISCONNECT THE
HARNESS PLUG TO RELAY
PACK "A." CHECK FOR
CONTINUITY FROM THE
INDIVIDUAL TERMINALS V
AND GY ON THE RELAY
PACK.

INFINITE
RESISTANCE

REPLACE
RELAY PACK "A"

RELAY
PACK

D/W
RUNS

VOLTAGE
CORRECT

ZERO
RESISTANCE

REPLACE
START RELAY

START RELAY

D/W
RUNS

VOLTAGE
CORRECT

REPLACE
RELAY PACK "B"

RELAY
PACK

D/W
RUNS

VOLTAGE
INCORRECT

REPLACE
M/C ASSY

M/C
ASSY

D/W
RUNS

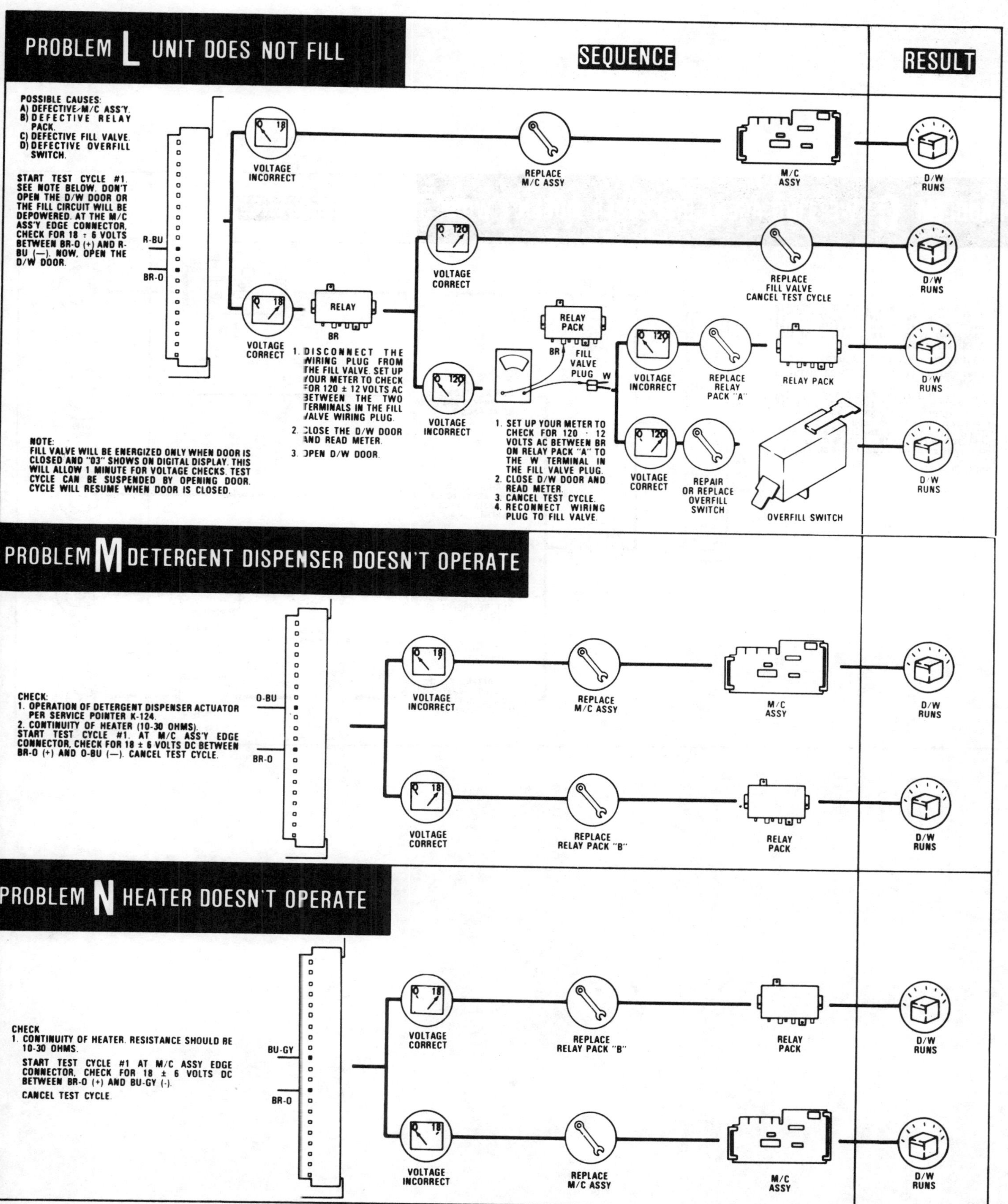

PROBLEM L UNIT DOES NOT FILL
SEQUENCE
RESULT
POSSIBLE CAUSES:
A) DEFECTIVE M/C ASS'Y.
B) DEFECTIVE RELAY PACK.
C) DEFECTIVE FILL VALVE.
D) DEFECTIVE OVERFILL SWITCH.
START TEST CYCLE #1. SEE NOTE BELOW. DON'T OPEN THE D/W DOOR OR THE FILL CIRCUIT WILL BE DEPOWERED. AT THE M/C ASS'Y EDGE CONNECTOR, CHECK FOR 18 ± 6 VOLTS BETWEEN BR-O (+) AND R-BU (−). NOW, OPEN THE D/W DOOR.
R-BU
BR-O
VOLTAGE INCORRECT
REPLACE M/C ASSY
M/C ASSY
D/W RUNS
VOLTAGE CORRECT
VOLTAGE CORRECT
RELAY
BR
1. DISCONNECT THE WIRING PLUG FROM THE FILL VALVE. SET UP YOUR METER TO CHECK FOR 120 ± 12 VOLTS AC BETWEEN THE TWO TERMINALS IN THE FILL VALVE WIRING PLUG.
2. CLOSE THE D/W DOOR AND READ METER.
3. OPEN D/W DOOR.
VOLTAGE INCORRECT
RELAY PACK
BR
FILL VALVE PLUG W
VOLTAGE INCORRECT
REPLACE RELAY PACK "A"
RELAY PACK
VOLTAGE CORRECT
REPAIR OR REPLACE OVERFILL SWITCH
OVERFILL SWITCH
1. SET UP YOUR METER TO CHECK FOR 120 · 12 VOLTS AC BETWEEN BR ON RELAY PACK "A" TO THE W TERMINAL IN THE FILL VALVE PLUG.
2. CLOSE D/W DOOR AND READ METER.
3. CANCEL TEST CYCLE.
4. RECONNECT WIRING PLUG TO FILL VALVE.
REPLACE FILL VALVE CANCEL TEST CYCLE
D/W RUNS
D/W RUNS
D/W RUNS
NOTE:
FILL VALVE WILL BE ENERGIZED ONLY WHEN DOOR IS CLOSED AND "03" SHOWS ON DIGITAL DISPLAY. THIS WILL ALLOW 1 MINUTE FOR VOLTAGE CHECKS. TEST CYCLE CAN BE SUSPENDED BY OPENING DOOR. CYCLE WILL RESUME WHEN DOOR IS CLOSED.
PROBLEM M DETERGENT DISPENSER DOESN'T OPERATE
CHECK:
1. OPERATION OF DETERGENT DISPENSER ACTUATOR PER SERVICE POINTER K-124.
2. CONTINUITY OF HEATER (10-30 OHMS).
START TEST CYCLE #1. AT M/C ASS'Y EDGE CONNECTOR, CHECK FOR 18 ± 6 VOLTS DC BETWEEN BR-O (+) AND O-BU (−). CANCEL TEST CYCLE.
O-BU
BR-O
VOLTAGE INCORRECT
REPLACE M/C ASSY
M/C ASSY
D/W RUNS
VOLTAGE CORRECT
REPLACE RELAY PACK "B"
RELAY PACK
D/W RUNS
PROBLEM N HEATER DOESN'T OPERATE
CHECK
1. CONTINUITY OF HEATER. RESISTANCE SHOULD BE 10-30 OHMS.
START TEST CYCLE #1 AT M/C ASSY EDGE CONNECTOR, CHECK FOR 18 ± 6 VOLTS DC BETWEEN BR-O (+) AND BU-GY (-).
CANCEL TEST CYCLE.
BU-GY
BR-O
VOLTAGE CORRECT
REPLACE RELAY PACK "B"
RELAY PACK
D/W RUNS
VOLTAGE INCORRECT
REPLACE M/C ASSY
M/C ASSY
D/W RUNS

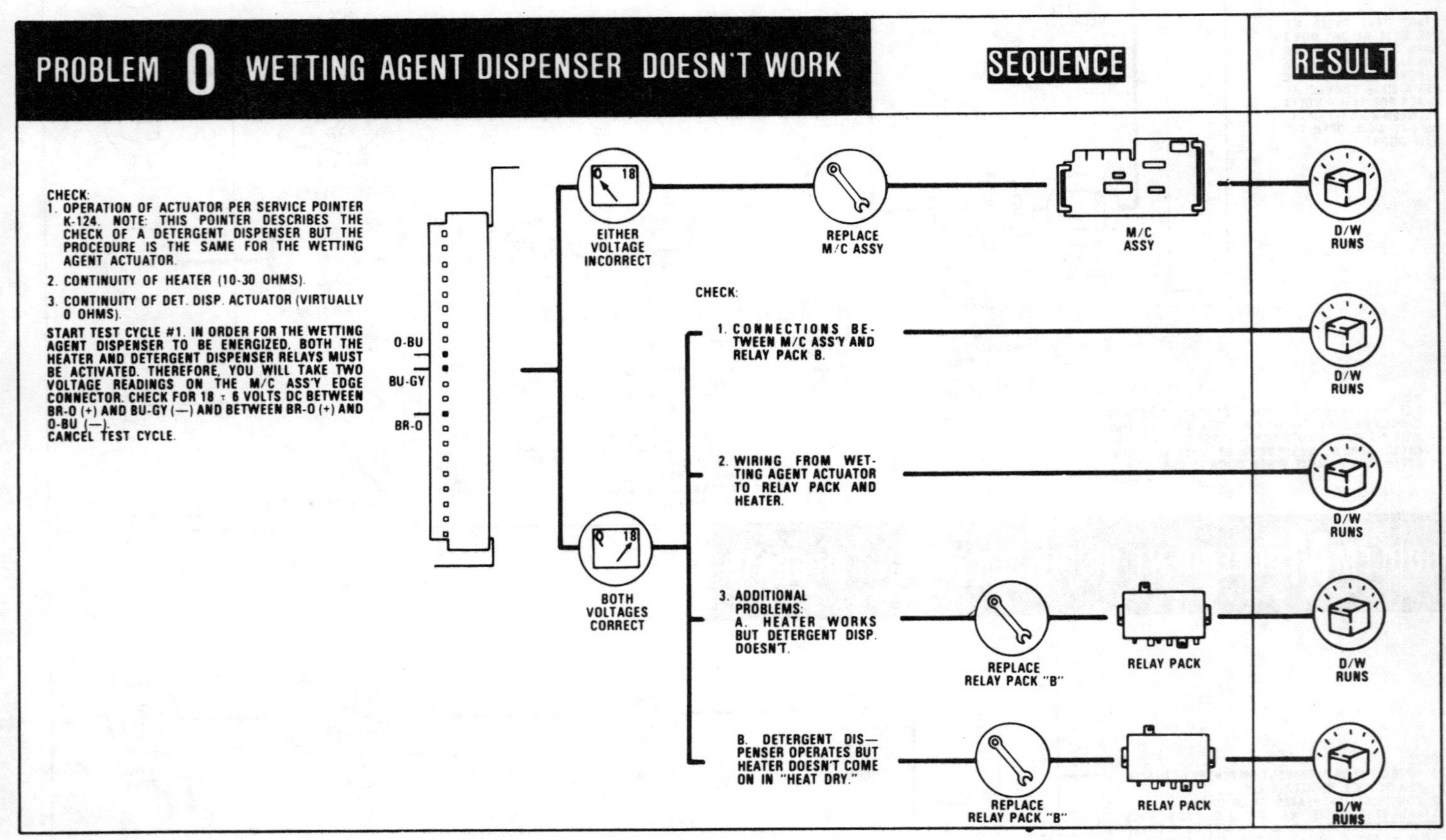
PROBLEM O WETTING AGENT DISPENSER DOESN'T WORK

SEQUENCE

RESULT

CHECK:
1. OPERATION OF ACTUATOR PER SERVICE POINTER K-124. NOTE: THIS POINTER DESCRIBES THE CHECK OF A DETERGENT DISPENSER BUT THE PROCEDURE IS THE SAME FOR THE WETTING AGENT ACTUATOR.

2. CONTINUITY OF HEATER (10-30 OHMS).

3. CONTINUITY OF DET. DISP. ACTUATOR (VIRTUALLY 0 OHMS).

START TEST CYCLE #1. IN ORDER FOR THE WETTING AGENT DISPENSER TO BE ENERGIZED, BOTH THE HEATER AND DETERGENT DISPENSER RELAYS MUST BE ACTIVATED. THEREFORE, YOU WILL TAKE TWO VOLTAGE READINGS ON THE M/C ASS'Y EDGE CONNECTOR. CHECK FOR 18 ÷ 6 VOLTS DC BETWEEN BR-O (+) AND BU-GY (—) AND BETWEEN BR-O (+) AND O-BU (—).
CANCEL TEST CYCLE.

O-BU

BU-GY

BR-O

EITHER
VOLTAGE
INCORRECT

REPLACE
M/C ASS'Y

M/C
ASS'Y

D/W
RUNS

CHECK:

1. CONNECTIONS BE-
TWEEN M/C ASS'Y AND
RELAY PACK B.

D/W
RUNS

2. WIRING FROM WET-
TING AGENT ACTUATOR
TO RELAY PACK AND
HEATER.

D/W
RUNS

BOTH
VOLTAGES
CORRECT

3. ADDITIONAL
PROBLEMS:
A. HEATER WORKS
BUT DETERGENT DISP.
DOESN'T.

REPLACE
RELAY PACK "B"

RELAY PACK

D/W
RUNS

B. DETERGENT DIS—
PENSER OPERATES BUT
HEATER DOESN'T COME
ON IN "HEAT DRY."

REPLACE
RELAY PACK "B"

RELAY PACK

D/W
RUNS

POWER SUPPLY

Whirlpool electric dishwashers require a two-wire, 1-phase, nominal 115-120 volt, 60 Hertz circuit. "Two-wire" is general terminology for a 115-120 volt system. Early (portable) dishwashers were equipped with a special three-wire power cord, Figure 2, of which the third wire (usually green in color) was used as an equipment ground (all dishwashers are so grounded), as a precaution against an electrical shock to the user. Such shocks may occur when an electrical current can travel to the metal frame or cabinet of the dishwasher from such causes as defective motor insulation, chaffed harness wiring touching adjacent metal parts, or by electrical components being water soaked.

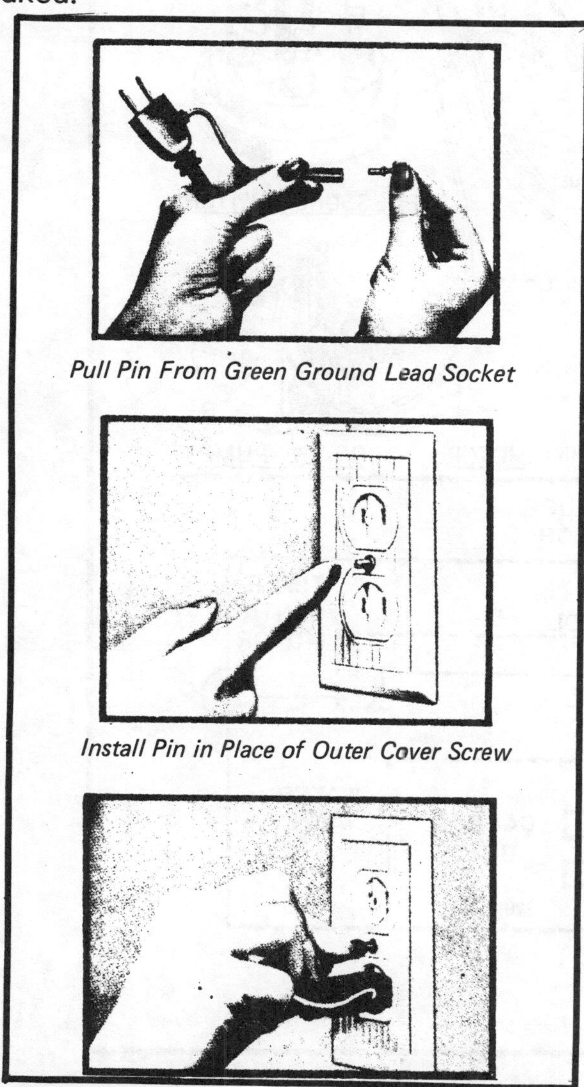

Pull Pin From Green Ground Lead Socket

Install Pin in Place of Outer Cover Screw

Plug in Power Cord and Slip Ground Lead Over Pin

Figure 2

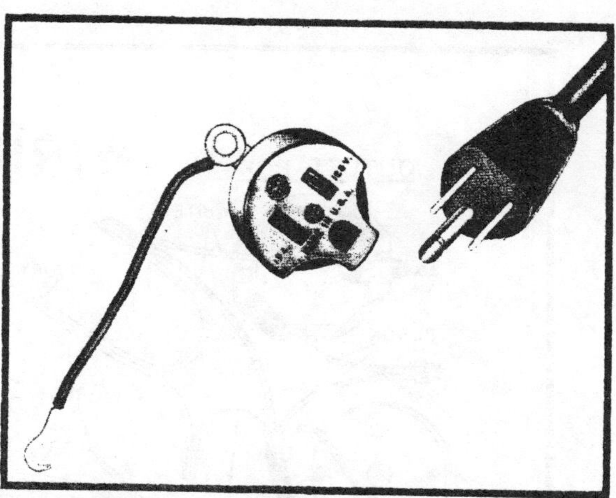

Figure 3 - Adapter and Three-Prong Plug

Later portable dishwashers are equipped with 3-prong plugs, Figure 3 together with the adapter shown. The latter is needed particularly, where the wall receptacle is the older type, as shown in Figure 4. and only when it is not feasible to install the proper grounded outlet. The green wire on the adapter is connected to the wall receptacle cover plate screw. It is essential that this method of grounding conforms to the local electrical code.

NOTE: Do not under any circumstances remove the ground prong from the power supply cord plug.

Figure 4 - Adapter Installed In Outlet

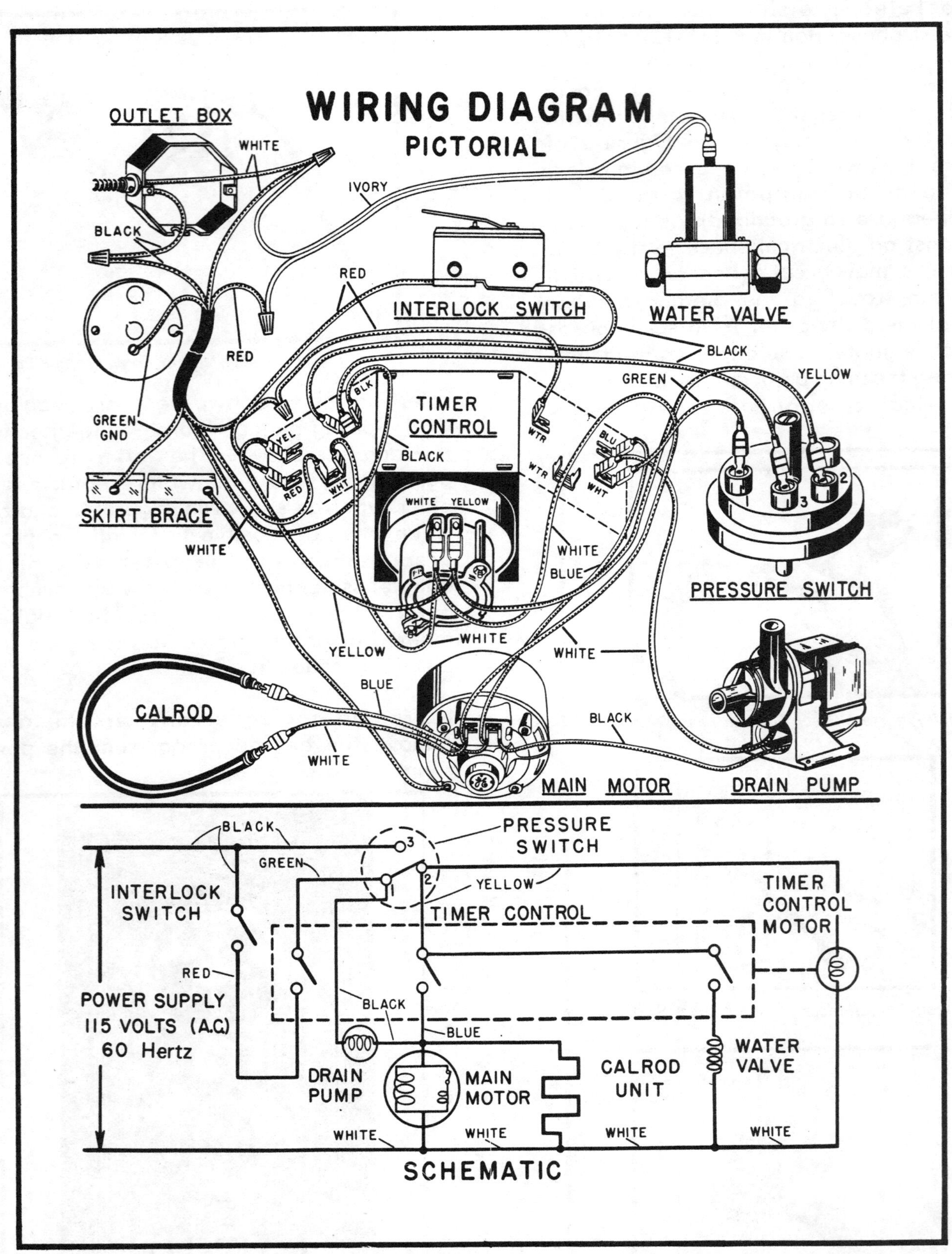

WIRING DIAGRAM
PICTORIAL
OUTLET BOX
WHITE
BLACK
IVORY
RED
INTERLOCK SWITCH
WATER VALVE
BLACK
RED
GREEN
YELLOW
BLK
TIMER
CONTROL
GREEN
GND
YEL
BLACK
WTR
BLU
RED
WHT
WTR
WHT
3
2
SKIRT BRACE
WHITE
WHITE
YELLOW
WHITE
WHITE
BLUE
PRESSURE SWITCH
YELLOW
WHITE
WHITE
CALROD
BLUE
BLACK
WHITE
MAIN MOTOR
DRAIN PUMP
BLACK
PRESSURE
SWITCH
GREEN
3
YELLOW
TIMER
CONTROL
MOTOR
INTERLOCK
SWITCH
1
2
TIMER CONTROL
RED
POWER SUPPLY
115 VOLTS (A.C.)
60 Hertz
BLACK
BLUE
WATER
VALVE
DRAIN
PUMP
MAIN
MOTOR
CALROD
UNIT
WHITE
WHITE
WHITE
WHITE
SCHEMATIC

*Figure 4*A

Free-standing and undercounter dishwashers are permanently wired-in to a 15 ampere branch circuit. These models require polarized electrical connections, as shown in Figure 4 A   Black lead is the "hot" side of the power line. Notice also, the green lead connected to the outlet box cover and to the dishwasher skirt brace. You must make certain the outlet box cover is firmly attached to the outlet box, otherwise, the dishwasher remains un-grounded. Thus at the time of the original installation, it is necessary to determine which of the two power leads is hot and which is "neutral." These then must be connected to the proper harness leads in the junction box.

A neon test light is useful in making the above determination. Touch a bare power lead with one test prod, with the fingers grasping the other prod. The hot lead will cause the neon bulb to glow. When the neutral lead is touched, the bulb will not glow. The polarity of the dishwasher should always be checked as a matter of routine whenever service is rendered to any of its electrical parts, unless you are absolutely certain that the polarity is correct.

Correct polarity means that the "hot" (black) wire from the power source is properly connected to the black wire in the dishwasher. The black "hot" wire from the power source should never be connected to the white wire in the dishwasher. In this case, the white wires in the dishwasher would become the "hot" wires, enabling the current to flow to all of the electrical parts in the dishwasher. The switches and fuse (or circuit-breaker) that control the dishwasher are located on the "hot" side of the circuit. Thus, if any of the electrical components should "short to ground" it could start unscheduled operation and cause considerable damage to the dishwasher.

Remove the appropriate panel and make the test described above on the black wire from the power cord, where it connects to the wire harness in the dishwasher outlet box. Do not forget to plug in the power cord. The heater

used in a Whirlpool dishwasher has an electrical rating in the approximate range of 13.5 amps, and from 600 to 800 watts, depending on the model. For precise information, refer to the nameplate or the service literature of the model in question.

ELECTRICAL TESTING

To check the continuity of a circuit in a wiring harness or of any component, it is best to use an externally powered continuity tester, such as the one shown in Figure 5.

This versatile continuity test cord can be a home-made affair that is used to test electrical current-carrying components. It may also be modified to serve as a live test cord or as a test lamp. As a live test cord, the lamp, Item 1, is replaced with a low amp fuse. This permits energizing the motors or solenoids with direct power, bypassing the machine wiring. As a test lamp, the male plug, Item 6, is shorted across its prongs. A shorted female connector works well as an adapter.When a continuity tester, which has been shunted to convert it to a test lamp, is used, you may check the existing power of current supplied to the dishwasher component.

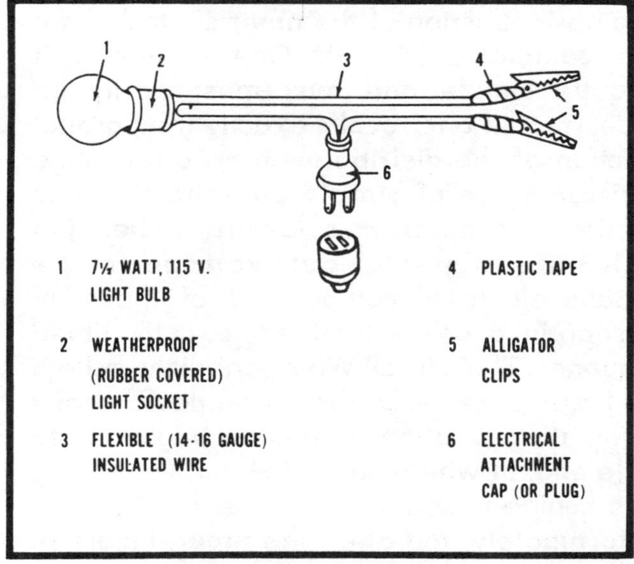

Figure 5

CAUTION: Always remove the service cord from the power supply or disconnect the circuit in some manner before using either an externally powered continuity tester or an ohmmeter.

Before making any tests, it will be to your
advantage to study the wiring diagram to
determine which harness leads to check for
continuity. Both ends of the wire or the
component part being checked should be
disconnected to insure against any possibility
of current feed-back through the machine
circuit. For example, let's say you are
checking the Calrod heating element. First,
disconnect the wires supplying current to the
heating element. Next, plug in the continuity
tester and apply the test clips of the tester to
the heating element terminals. If the test
lamp lights, proper continuity is indicated.
Now, let's say the wires supplying current to
the element are NOT disconnected and we
apply the test clips to the element terminals.
Again the test lamp will light, but we haven't
positively determined whether or not the
element is good, because of the possibility of
current "feed-back" through the machine
wiring.

NOTE: Unless otherwise stated, all electrical
tests are made with the timer turned to ON.
Be sure the power is disconnected before any
disassembly is attempted.

TIMERS

The basic function of the timer control is not
at all complex or intricate. On the contrary, it
is quite simple and may most easily be
understood if one recalls exactly the parallel
function of the distributor in an automobile
ignition system. Simply put, the timer is
another distributor, its sole purpose being to
distribute or parcel out, voltage to the
various electrical components of the dishwasher in a predetermined, exactly timed
sequence. Thus, in all Whirlpool dishwashers
the timer, once set by the user, allows wholly
automatic operation from cycle beginning to
cycle end, at which point it shuts off, halting
all machine functions until reset by the user.
Unfortunately, too often the timer, Figure 6,
is the "scapegoat" for many problems which
are in fact caused by the malfunction of some
other component. For this reason a timer
should never be condemned until all other
possibilities have been eliminated and a sure
test, or tests, prove timer malfunction.

The timers used on all RCA WHIRLPOOL
Dishwashers are similar in design and
construction. Once the timer is activated, it
will control electrical circuits as required in
the cycle to provide full automatic function.
Timers will vary slightly due to the number of
electrical components used on the different
models.

The timer assembly consists of three basic
components — timer motor, escapement,
switch box and cam assembly. The drive gear
of the timer motor actuates the escapement
which in turn advances the cam at 45-second
intervals. Cam rotation opens and closes
timer switch contacts to provide function of
electrical components. The timer motor is
operated continuously throughout the machine cycle.

All timers are equipped with a master line
switch. Its function is to provide a means of
opening the line circuit while setting the
timer in the cycle and to provide a means of
interrupting the cycle once the timer has
been set.

Figure 6

A later refinement in timer design, to all intents eliminated the escapement mechanism as such, it having been replaced by a direct motor drive. Timer motors, Figure 7, will vary in size and appearance due to the various sources of manufacture, but these differences not withstanding, all function in the same manner. The mechanical power of the timer motor is transmitted by means of a small, slowly rotating shaft and pinion, which drives a gear in the escapement, if so equipped.

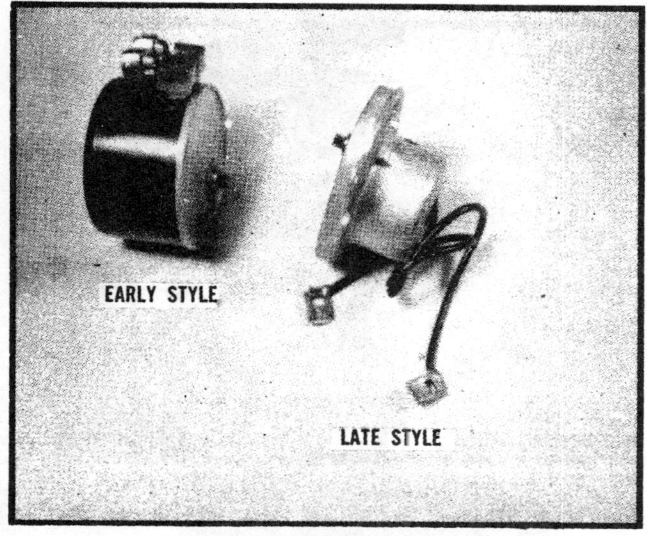

Figure 7 - Two Styles of Timer Motors

In the "impulse" timer, the escapement is a spring-powered clockwork mechanism that advances the timer cam shaft a set number of degrees at every periodic advance. Both the degree of advance and the timed interval of advance depend on the particular design of the timer.

The timer motor winds up a music wire torsion spring to a predetermined tension. After the required time lapse, the spring unwinds abruptly, causing a chain of gears to advance the timer cam shaft the predetermined number of degrees, causing the audible "click," characteristic of the impulse timer.

The multiple circuit cam switches, Figure 8, ride on cams which cause contacts to open and close as the brass arms are raised and lowered by the cam profile. Since the cam shaft is quickly advanced in short steps by the action of the escapement, the switch contacts open and close with a snap action intended to reduce arcing damage, a necessary feature common to all switches carrying more than negligible current.

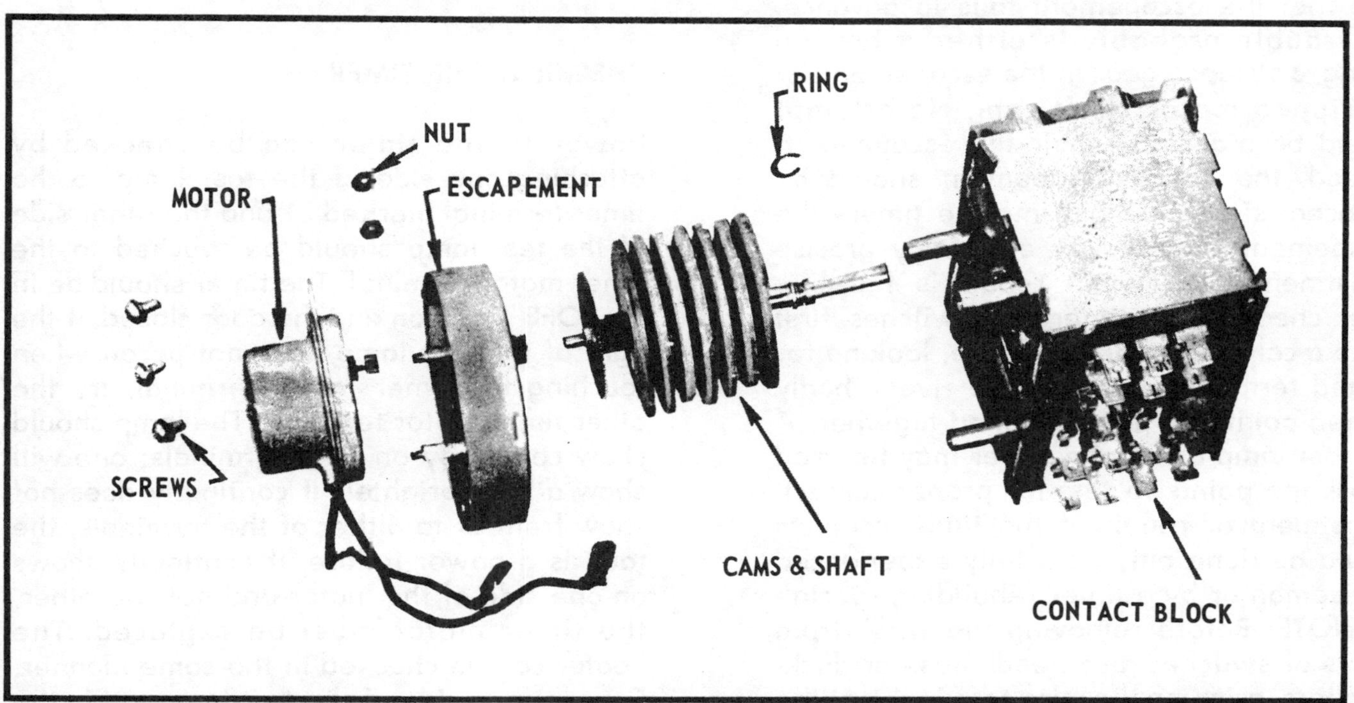

Figure 8 - Switch Controls Exposed on Timer Control

Coin silver contact points are imbedded in the ends of the spring brass switch arms. Silver is a favorite contact material because satisfactory electrical connection is made even through blackened and pitted contact points. The timer dial or knob is manually rotated from the control panel, but (on all "impulse" timers) must be turned only in a clockwise direction. If it is turned counter-clockwise, the cam indents will jam against and damage the contact arms. Damage might also be done to the escapement, since it is designed to rachet in one direction only.

Testing a timer is a combination of observing, listening, and checking electrical continuity. In quiet surroundings the faint whirring sound of the timer motor sometimes can be heard, indicating that it is probably operating satisfactorily. The motor may be removed from the timer and energized by the use of the test cord, so that the shaft rotation may be observed from positive indication that it is working. Take note that a defective timer motor may be replaced without replacing the entire timer. In the impulse timers a properly working escapement will advance the timer at evenly spaced intervals. When the timer motor pinion is rotating properly and you note that the escapement fails to advance, the trouble probably is either a broken spring, a stripped gear in the escapement, or a stripped motor shaft cam. No attempt should be made to repair the escapement; instead, the entire escapement should be replaced, since on most impulse timers the escapement requires an extremely precise adjustment even when repair is feasible. When checking the timer cam switches, first make a careful visual inspection, looking for burned terminal boards, loose rivets, badly burned points, or points welded together. A live test lamp or an ohmmeter may be used across the points to test for proper contact. Adjustment or repair of the timer contacts should be done only by a fully experienced serviceman or by a timer rebuilding specialist. NOTE: Before removing the wires from timers or switches, use needlenose or duck-bill pliers, gripping the wire terminal--not the wire, or better still, pry off the terminals with

a screwdriver as shown in Figure 9. Do not carelessly yank on the wires; you may break them off the terminals. Because the timer is a precision instrument and can easily be maladjusted, we strongly recommend that this unit not be disassembled in the field, other than removing or replacing the timer motor. The test lamp, Figure 5, may be used to check the timer in place. The described tests may be followed by referring to the wiring diagrams.

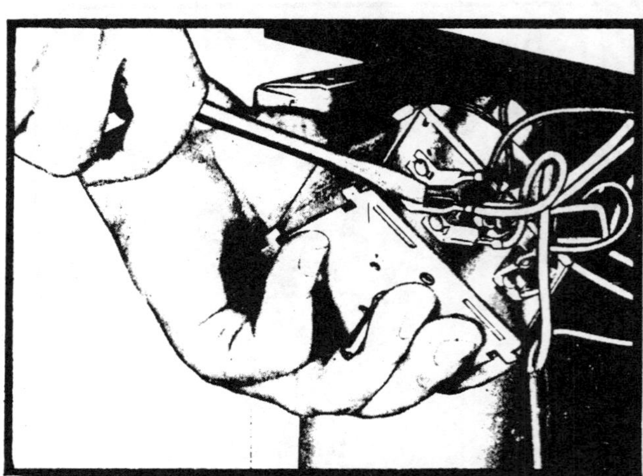

Figure 9

CHECKING THE TIMER

Power to the timer can be checked by attaching one side of the test lamp to the timer terminal marked L1 and the other side of the test lamp should be touched to the timer motor terminal. The timer should be in the "OFF" position and the door closed. If the light of the test lamp does not go on when touching the timer motor terminal, try the other timer motor terminal. The lamp should show continuity on both terminals; one will show a little brighter. If continuity does not show from L1 to either of the terminals, the fault is a power failure. If continuity shows on one side of the motor and not the other, the timer motor must be replaced. The heater can be checked in the same manner. Connect one side of the test lamp to L1, the other side to the heater terminals. If the test lamp does not light on either heater terminal you have a power failure; if it lights on one side of the heater and not the other, the heater element is bad and must be changed; if it lights bright from one terminal and a dimmer light when moving the test probe over, then you have power and the heating element is operative. The fill valve can be checked the same way. Remember that the timer should be on the "OFF" position when this test is made. If the customer's complaint is any of the following:

(a) Timer fails to advance
(b) Heater does not go on
(c) Water does not enter dishwasher

If you have completed the above test and both sides of the components showed continuity, then through a series of elimination it would indicate the timer is at fault and must be replaced.

TIMER CHECK

Before attempting to remove timer for replacement, the following tests should be made.

1. All other components that are controlled by the timer should be tested.

2. Take note if timer advances. Allow at least two minutes for this test.

3. If timer knob fails to advance, remove motor from the timer and test. See text "Testing the Timer Motor" also "Checking the Escapement" if the timer you are servicing has a separate escapement.

4. If motor operates and timer does not advance, it would indicate internal defects in the timer assembly and timer must be replaced. Make a diagram of the timer wiring if a schematic is not available, before disconnecting any wires.

TIMER REMOVAL

1. Remove the rear shroud, see text "Rear Shroud."

2. Loosen the clamp screw under the rubber protector cup. Slide the parts upward and disengage timer extension shaft from the timer shaft, Figure 10.

3. Remove the timer mounting screws and push timer forward to disconnect the timer leads.

4. When reinstalling timer, double check the wiring before assembling the shroud.

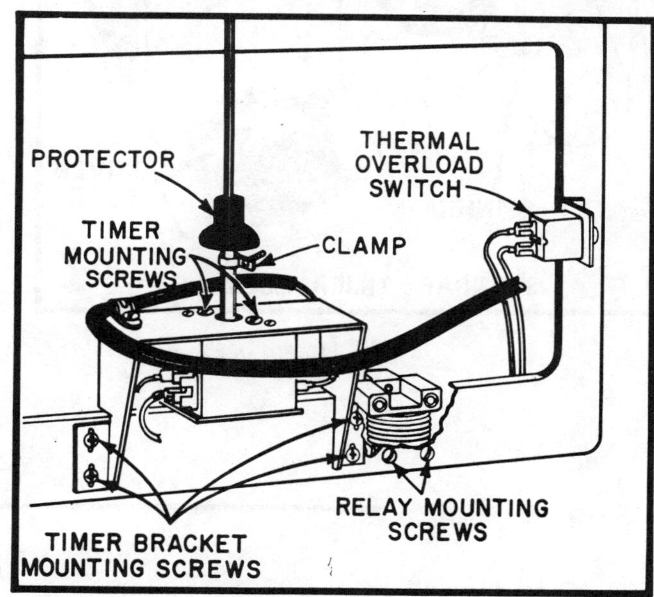

Figure 10

TIMERS, Late and Current Models

The timers used on Whirlpool Automatic Dishwashers are very similar in basic principle and design. They differ in physical appearance due to manufacturers design and complexity. This enables various functions and features incorporated in the numerous models. They all have a synchronous type motor similar to those used in electric clocks. In replacing the motor, care should be taken that the motor not only physically fits the mounting, but that the gear and pinion are positioned correctly, and that the voltage, number of teeth in the gear, length of pinion, the RPM's, and direction is the same when using a replacement motor, rather than the OEM factory replacement part, see Figure 11.

The timers used on 1970 models thru the current models are all pull-to-start with the exception of the rapid advance systems. These are controlled by a push switch that is mounted on the dishwasher console. The only replaceable component or part on the timer is the timer motor, see text "TIMER MOTOR." The rapid advance timer has a replaceable RA motor in addition to the timer motor, Figure 12.

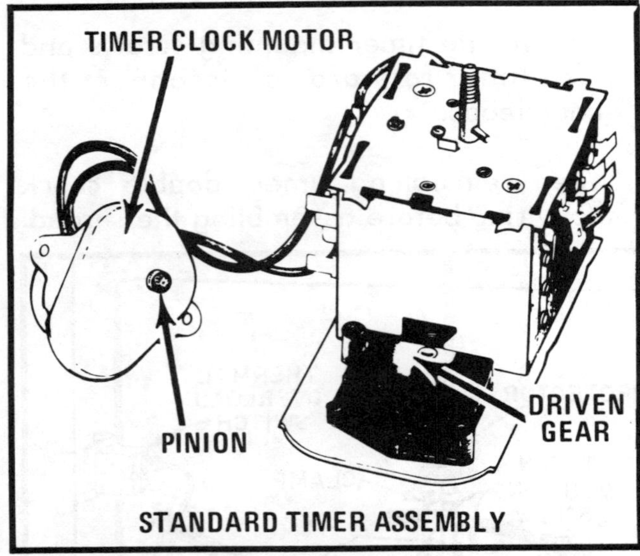

Figure 11

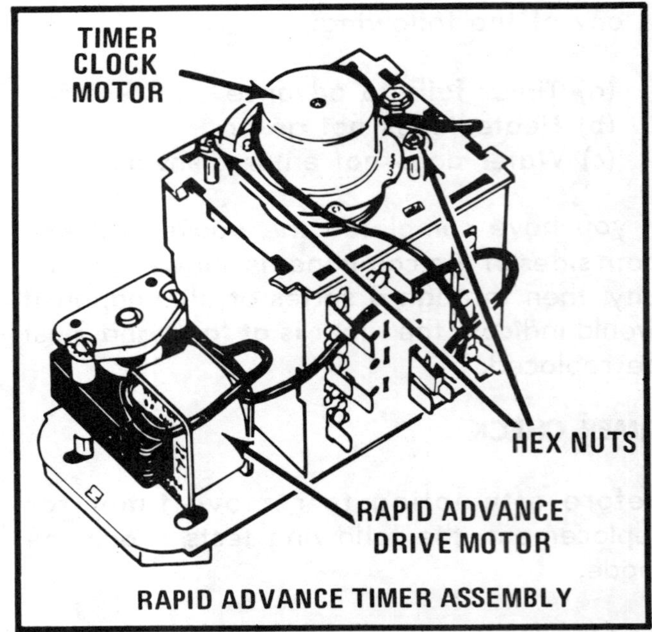

Figure 12

QUICK DISCONNECT TIMERS, Current Models

With the exception of Model SDU 3000-0 all models have a quick disconnect timer. These are similar to the quick disconnect timers used on Whirlpool Automatic Washing Machines currently. The wire harness plugs into the timer as shown in *Figure 12A.* This will preclude any wiring error on the part of the service technician, and is also a time saver.

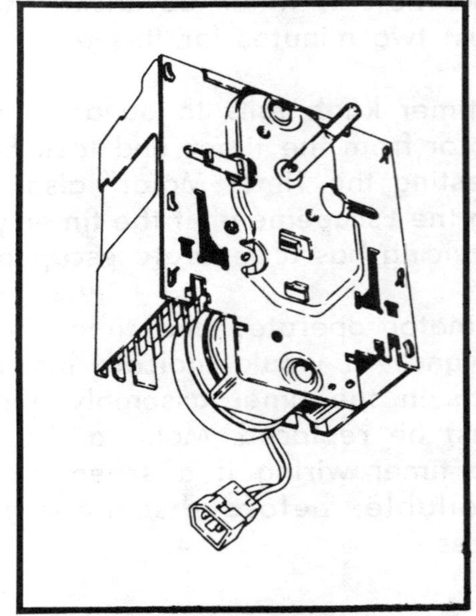

Figure 12A

The rapid advance drive motor advances the timer cams to a pre-determined starting position. This starting position is programmed through the push button selector switch and various contacts in the timer itself for any given selection. This is similar to the "Short Cycle" as advanced manually on the standard timer. The table below, Figure 13, shows how the timer advances through the cycle for any given selection.

The 60th increment is the stop position of the cycle. The contact "X" in the pushbutton switch momentarily is closed at the bottom of all pushbutton excursions, with the excep-

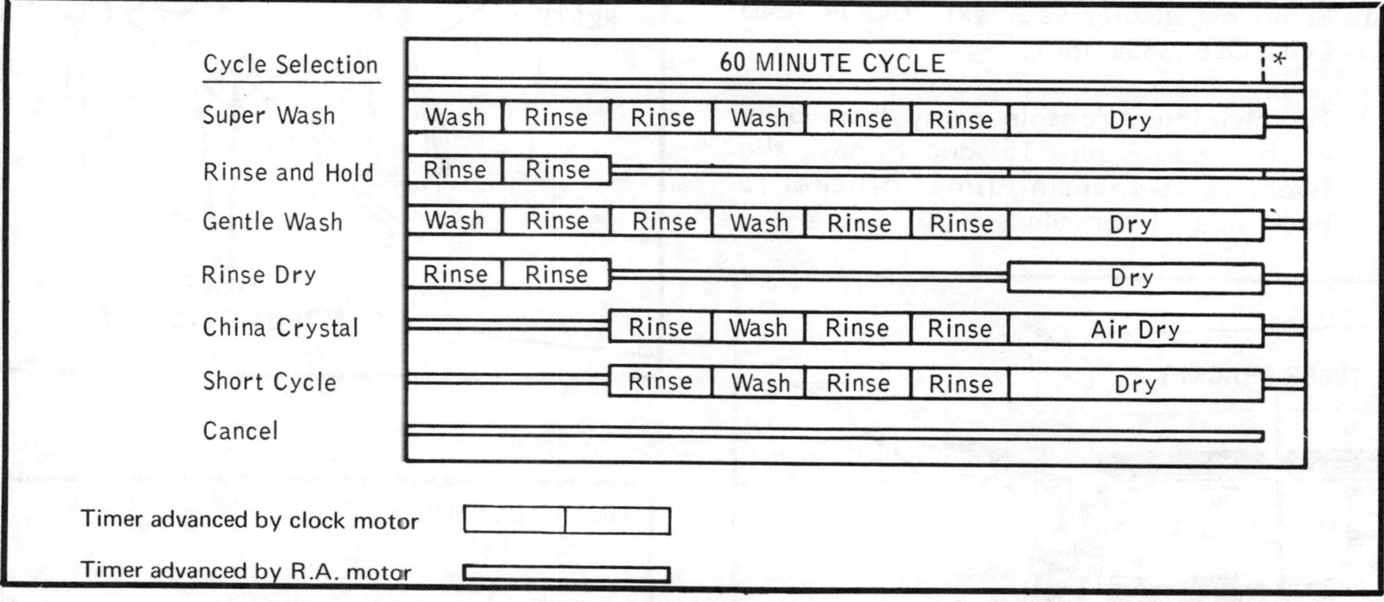

Figure 13

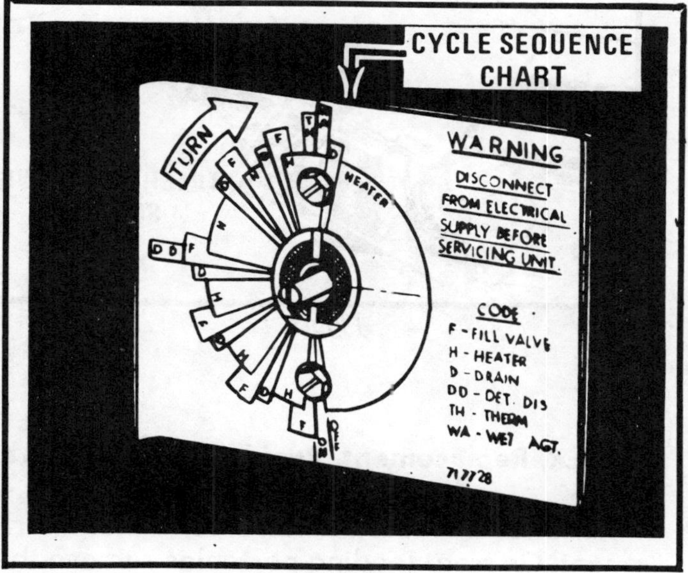

Figure 14

tion of "cancel." The "X" contact completes the circuit to the Rapid Advance motor, advancing the timer back to the first increment, initiating a new cycle.

NOTE: A sequence chart is attached to the timer mounting bracket, Figure 14, to be used for trouble diagnosis. There is a slot provided in the end of the shaft so the timer can be advanced manually with the use of a screw driver. The flat of the shaft serves as the dial pointer. See "TIMER CHECK" text.

TIMER, Replacement, Late and Current Models.

The following text does not refer to Rapid Advance timers.

1. Remove console, see text "DOOR AND CONSOLE ASSEMBLY."

2. Tilt aluminum console away from dishwasher, see Figure 15, and remove the two screws securing timer bracket to back side of console.

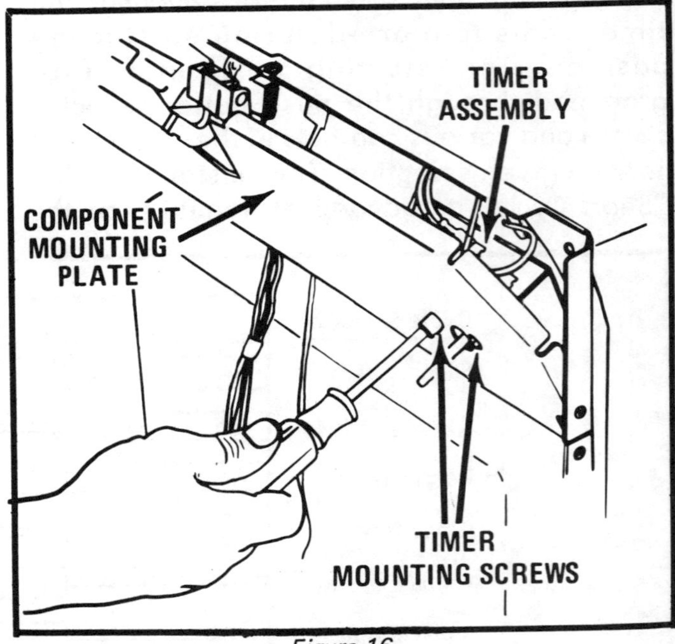

Figure 16

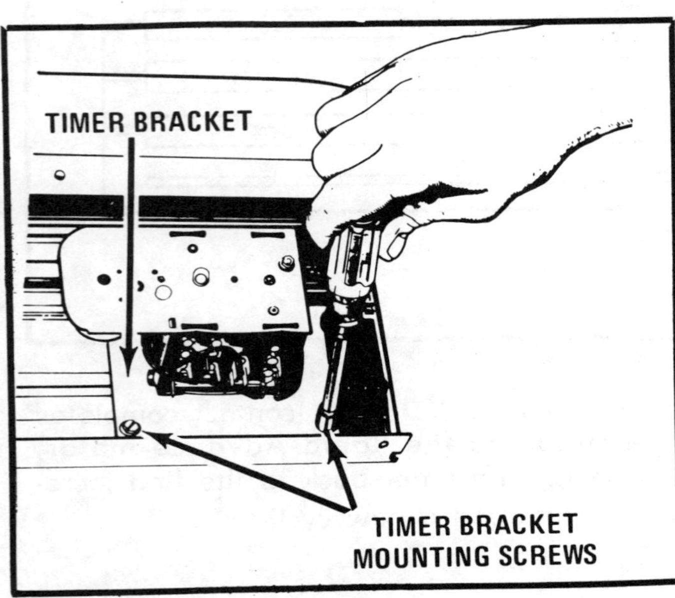

Figure 15

Figure 17

3. Remove two screws that hold the timer to the mounting bracket, Figures 16 and 17.

4. Remove leads from all timer terminals, make a chart if schematic is not available.

5. To reinstall timer, reverse procedure. Care should be taken that the wiring is properly connected.

TIMER, Replacement, Rapid Advance Timers.

The timer is located in the machine compartment. On front loading portables remove the descending access panel.

1. Remove access cover assembly secured with four screws.

2. Remove the two screws securing timer to cover assembly, Figure 18.

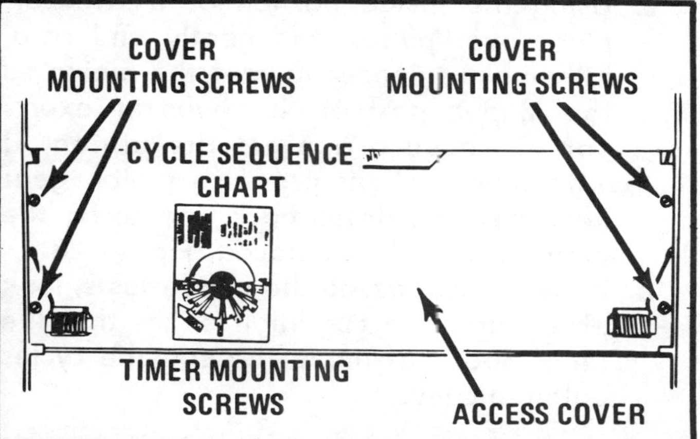

Figure 18

3. On undercounter models remove four screws from toe plate-access panel assembly.

4. Remove two screws securing timer mounting bracket fastened to left leg of washer, see Figure 19.

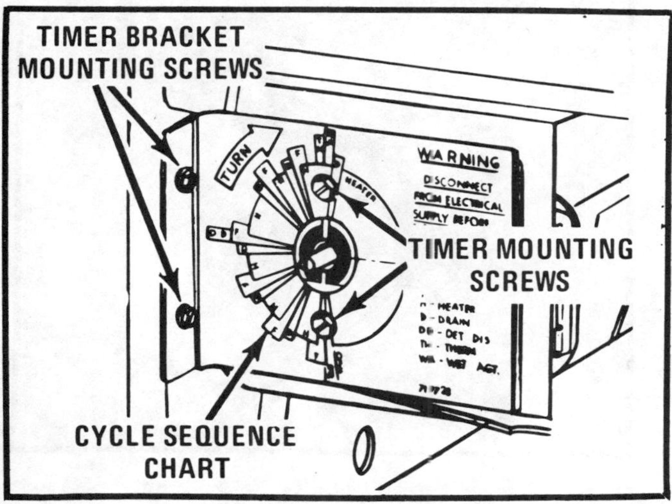

Figure 19

5. Remove two screws securing timer to bracket.

6. Remove all wiring connections from timer, if a schematic is not available, make a chart.

7. Remove timer from dishwasher, to reinstall timer, reverse procedure.

To remove the timer motor, see text "TIMER MOTOR," Removal, Figure 20.

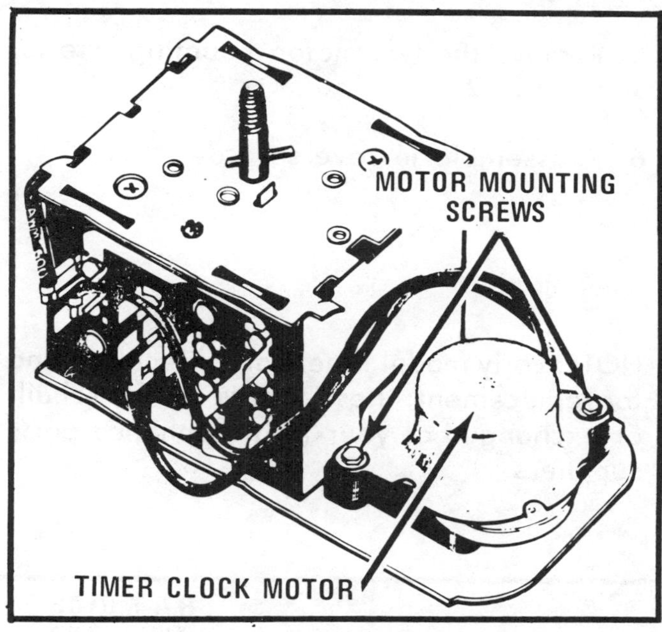

Figure 20

TIMER MOTOR, R.A. Motor, Replacement.

1. Remove timer from dishwasher, see above text.

2. Remove cover from opposite end of timer, Figure 21, bend locking lugs to remove cover.

3. Remove gear train cover, Figure 21, bend locking lug to remove.

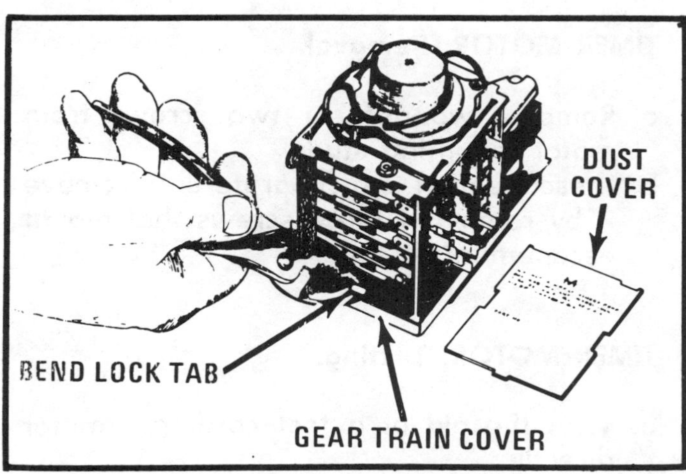

Figure 21

4. Remove R.A. motor leads from timer terminals.

5. Remove the two motor mounting screws, Figure 22.

6. Reassemble in reverse order.

NOTE: Early model timers are difficult to find for replacement; these timers can be rebuilt or exchanged at your local appliance parts suppliers.

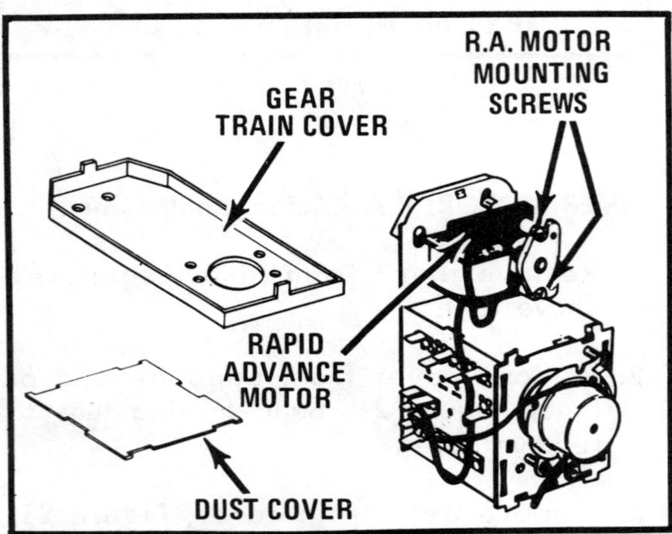

Figure 22

TIMER MOTOR, Removal.

a. Remove two nuts or two screws from motor mounting studs.
b. If escapement is a separate unit, remove it by removing three screws that mount escapement to timer.

TIMER MOTOR, Testing.

a. With the aid of a test cord, run motor directly.

b. If pinion gear hesitates or rocks back and forth, motor should be replaced.
c. Using the inside portion of a tweezer, place over pinion and gently and carefully move tweezer toward the pinion so the hinged part of the tweezer exerts some pressure against the pinion. If with only a slight pressure pinion gear hesitates or rotates back and forth, the motor should be replaced, Figure 23.
d. If the motor passed the above tests, look elsewhere, if a condition exists that the timer does not move through the cycles automatically.

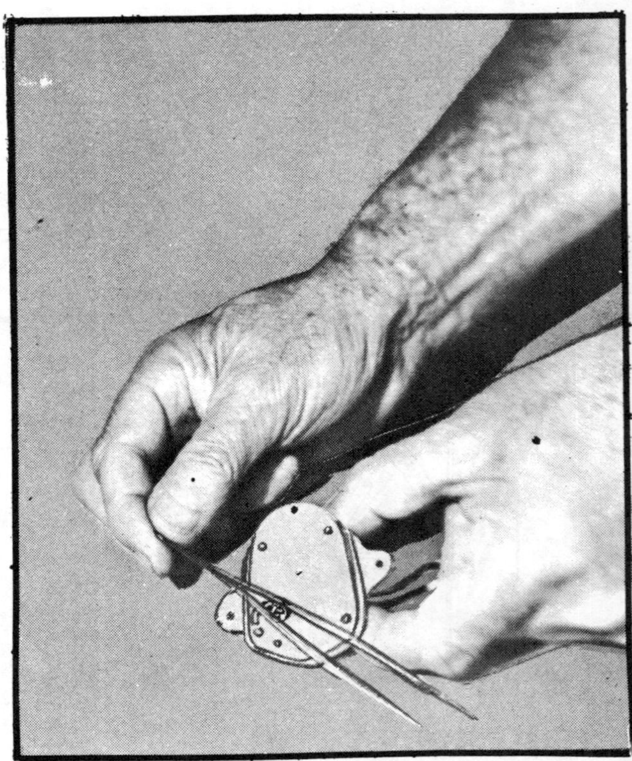

Figure 23

ESCAPEMENT, Testing.

NOTE: Older models had a separate escapement, Figure 24. This type of unit would "wind up" and at a predetermined time, the slide retainer would allow the gear to "unwind." Each of these wind and unwind periods are "Increments."

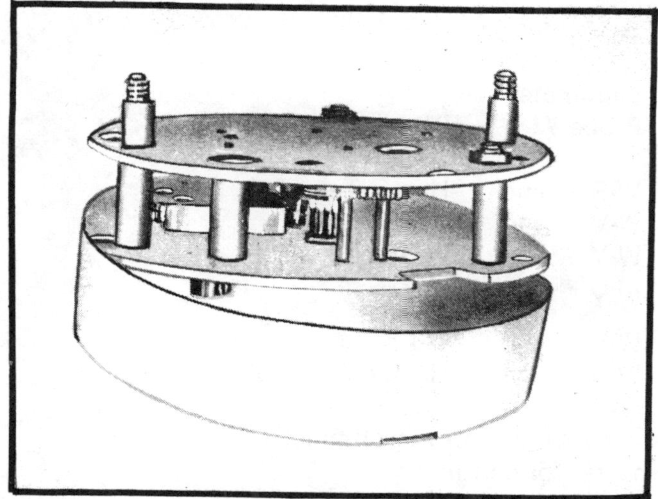

Figure 24

Replace timer motor back on escapement if it has been removed. Connect motor to a direct cord. Proceed as follows:

a. Remove cover from escapement by turning until grooves line up with the indents on the cover, then lift off.
b. Place tweezers over pinion as previously outlined, Figure 24A.
c. Wait for at least one minute for timer to act.
d. If pinion gear fails to move, escapement must be replaced.
e. The old style motor, as described above, can be replaced with the new type motor.

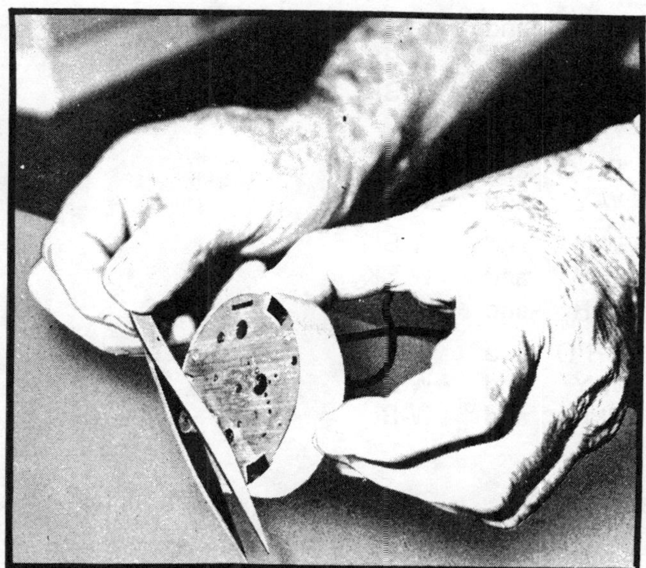

Figure 24A

PUSHBUTTON SWITCH

The cycle selection is made by selecting one or a combination of electric circuits through the switch contacts. The switch may have from three to eight buttons, depending on the features of a particular model. The pushbutton switch is located in the console and access to the switch is gained by removing console panel, see "DOOR AND CONSOLE ASSEMBLY" text.

PUSHBUTTON SWITCH, Testing.

Disconnect power source.

1. Remove console assembly.

2. Tip console assembly as shown in Figure 25.

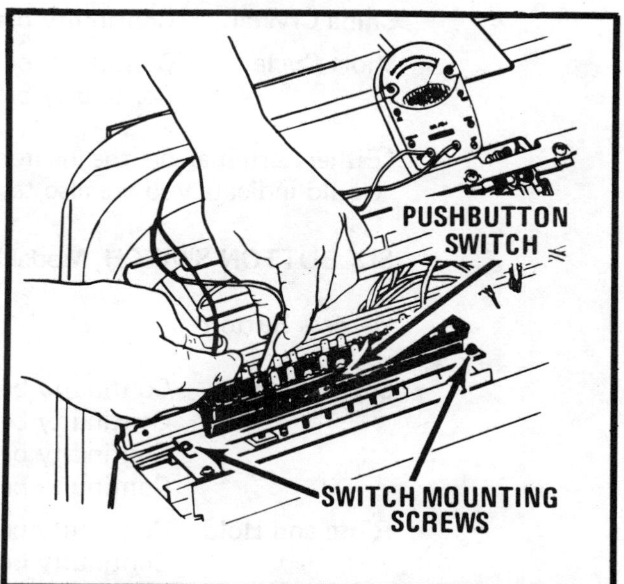

Figure 25

3. Remove wire leads from switch terminals, if a schematic is not available, make a wiring chart.

4. With the aid of a ohmmeter check continuity as follows.

THREE BUTTON SWITCH

Depress Button		Ohmmeter Probe #1		Probe #2
Super Wash	Continuity between	W-R	and	*Heater Terminal
	Continuity between	W-V	and	W-BK
	Continuity between	W-Y	and	W-BK
Rinse and Hold	Continuity between	W-Y	and	W-BK
Short Cycle	Continuity between	W-V	and	W-BK
	Continuity between	W-R	and	*Heater Terminal

*Either terminal on the heater, a resistance shown on the ohmmeter would indicate you are also testing the heater for continuity.

FOUR BUTTON SWITCH

Depress Button		Ohmmeter Probe #1		Probe #2
Super Wash	Continuity between	W-Y	and	W-BK
	Continuity between	W-V	and	W-BK
	Continuity between	W-R	and	*Heater Terminal
Rinse and Hold	Continuity between	W-Y	and	W-BK
China Crystal	Continuity between	W-V	and	W-BK
Short Cycle	Continuity between	W-V	and	W-BK
	Continuity between	W-R	and	W-BK

*Either terminal on the heater, a resistance shown on the ohmmeter would indicate you are also testing the heater for continuity.

SIX BUTTON SWITCH, Model STP 90-3

Depress Button		Ohmmeter Probe #1		Probe #2
Super Wash	Continuity between	BR-O	and	BR
	Continuity between	W-Y	and	W-BK
	Continuity between	W-V	and	W-BK
	Continuity between	W-R	and	W-R
Rinse and Hold	Continuity between	BR-O	and	BR
	Continuity between	W-Y	and	W-BK
Gentle Wash	Continuity between	W-Y	and	W-BK
	Continuity between	W-V	and	W-BK
	Continuity between	W-R	and	W-R
Rinse and Dry	Continuity between	BR-O	and	BR
	Continuity between	W-Y	and	W-BK
	Continuity between	W-R	and	W-R

SIX BUTTON SWITCH Model STP 90-3 (Continued)

China Crystal	Continuity between	W-V	and	W-BK
	Important Note*	*W-R	and	W-R
Short Cycle	Continuity between	BR-O	and	BR
	Continuity between	W-V	and	W-BK
	Continuity between	W-R	and	W-R

*Important Note: On Model SWP 90-0, there should be NO CONTINUITY between W-R and W-R.

SIX BUTTON SWITCH

Depress Button		Ohmmeter Probe #1		Probe #2
Super Wash	Continuity between	W-Y	and	W-BK
	Continuity between	W-V	and	W-BK
	Continuity between	W-R	and	*Heater Terminal
Rinse and Hold	Continuity between	W-Y	and	W-BK
Gentle Wash	Continuity between	W-T	and	W-T
	Continuity between	W-Y	and	W-BK
	Continuity between	W-V	and	W-BK
	Continuity between	W-R	and	*Heater Terminal
Rinse and Dry	Continuity between	W-Y	and	W-BK
	Continuity between	W-R	and	*Heater Terminal
China Crystal	Continuity between	W-T	and	W-T
	Continuity between	W-V	and	W-BK
	Continuity between	W-R	and	W-R
Short Cycle	Continuity between	W-V	and	W-BK
	Continuity between	W-R	and	*Heater Terminal

*Either terminal on the heater, a resistance shown on the ohmmeter would indicate you are also testing the heater for continuity.

SEVEN BUTTON SWITCH

Depress Button		Ohmmeter Probe #1		Probe #2
Super Wash	Continuity between	W-Y	and	W-BK
	Continuity between	W-BU	and	W-BK
	Continuity between	W-V	and	W-BK
	Continuity between	W-R	and	W-R (Heater)
Except on Model STP 100-2		BR-O	and	BR
Rinse and Hold	Continuity between	W-Y	and	W-BK
	Continuity between	W-BU	and	GY-P
	Continuity between	W-V	and	GY-P
Except on Model STP 100-2		BR-O	and	BR

SEVEN BUTTON SWITCH   (Continued)

Gentle Wash	Continuity between	W-Y	and	W-BK	
	Continuity between	W-BU	and	W-BK	
	Continuity between	W-V	and	W-BK	
	Continuity between	W-BK	and	T1	
Except Models SVP 100-0		W-T	and	W-T(Air V.)	
STP 100-3		W-R	and	W-R(Heater)	
STP 100-4					
Rinse and Dry	Continuity between	W-Y	and	W-BK	
	Continuity between	W-BU	and	W-BK	
	Continuity between	W-V	and	GY-P	
	Continuity between	W-R	and	W-R(Heater)	
Except Models STP 100-0		BR-O	and	BR	
STP 100-2					
China Crystal	Continuity between	W-Y	and	GY-P	
	Continuity between	W-BU	and	W-BK	
	Continuity between	W-V	and	W-BK	
	Continuity between	W-BK	and	T1	
Except Models SVP 100-0		W-T	and	W-T(Air V.)	
STP 100-3					
STP 100-4					
	Continuity between	W-R	and	W-R(Heater)	
Short Cycle	Continuity between	W-Y	and	GY-P	
	Continuity between	W-BU	and	W-BK	
	Continuity between	W-V	and	W-BK	
	Continuity between	W-R	and	W-R(Heater)	
Except Models STP 100-0		BR-O	and	BR	
STP 100-2					
Cancel	Continuity between	W-Y	and	GY-P	
	Continuity between	W-BU	and	GY-P	
	Continuity between	W-V	and	GY-P	

DRY SELECTOR SWITCH

All SDU and SDF model dishwashers feature a Dry
Selector switch mounted in the console, *Figure 25A;*
On all models but the SDU 9000 model, this switch
is a rocker type, two position switch. The user has
the option of "air dry" or "heat dry". The switch
is wired in series with the heater element and as such
when used as "air dry" will use much less energy.

CANCEL/DRAIN, SDU 9000

The pushbutton switch on the SDU 9000 has a Can-
cel/Drain selection. This selection rapid advances
the timer to the final drain position. The period of
drain will be approximately two minutes. Timer
will then rapid advance to the off position.

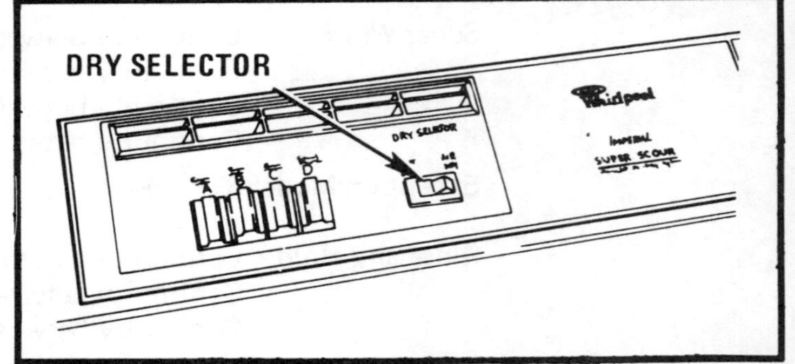

Figure 25A

WATER PRESSURE SWITCH

The water pressure switch is used as a safety switch to prevent overflowing of the water in the dishwasher. Most of the Whirlpool dishwashers use a water pressure switch, with the exception of two models. Although the water pressure switch will control the electric circuit to the water fill valve solenoid and open the circuit to the solenoid when the dishwasher overfills, it will not control the water valve if the water valve is mechanically stuck. See text, "WATER VALVE." On a time-fill dishwasher, where the water fill is calibrated to the water pressure, the safety feature of the water valve will prevent an overfill in case of excessive water pressure. If the problem is that water will not enter the dishwasher, and the water pressure switch is suspected, it can be tested as follows:

The water pressure switch is in series with the water fill valve solenoid.

1. Shut off the power to the dishwasher.

2. Remove the two wires from the water pressure switch, Figure 26.

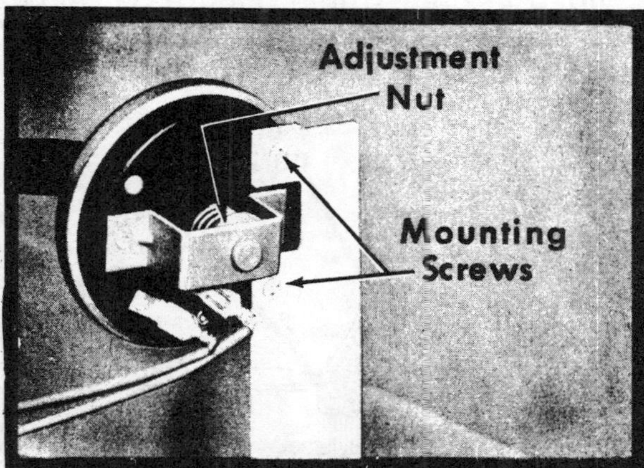

Figure 26

3. By twisting the wires in opposite directions, the spade connectors can be fastened together by slipping one edge of the spade connector into the edge of the other. This will complete a temporary by-pass.

4. Reconnect the power, close lid or door and latch.

5. Turn control to fill. If water comes into the dishwasher it would indicate the water pressure switch is defective and must be replaced.

6. If the complaint is that water keeps coming into the machine, the valve could be mechanically stuck open. See text, "WATER VALVE."

On later models the pressure switch has a twofold purpose. It is a single pole, double-throw switch. The normally closed contacts are in series with the water fill valve solenoid and the normally open contacts are in circuit between L1 and the drain pump and L2. In case of overfill the safety will trip and in so doing the drain pump will start. When the proper water level is reached, the water pressure switch will return to the normally closed contact position. As such, it truly acts as a safety valve.

DOOR AND CONSOLE ASSEMBLY, Models With Extruded Aluminum Consoles, Disassembly.

1. Disconnect power source.

2. Remove descending access panel on front loading portable models.

3. Remove the two screws securing each end cap to console, Figure 27.

Figure 27

4. Slide door side trim pieces up and off of door frame, Figure 28.

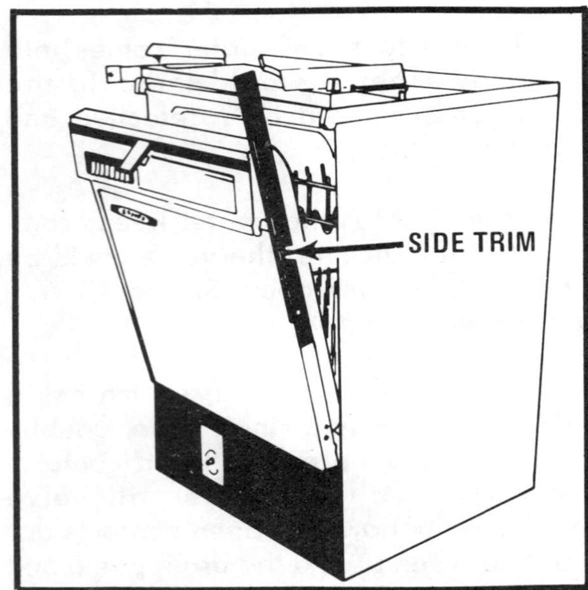

Figure 28

5. Pull out on bottom of front panel and disengage from groove in the console.

6. Remove handle from the door latch mechanism, remove screw, Figure 29.

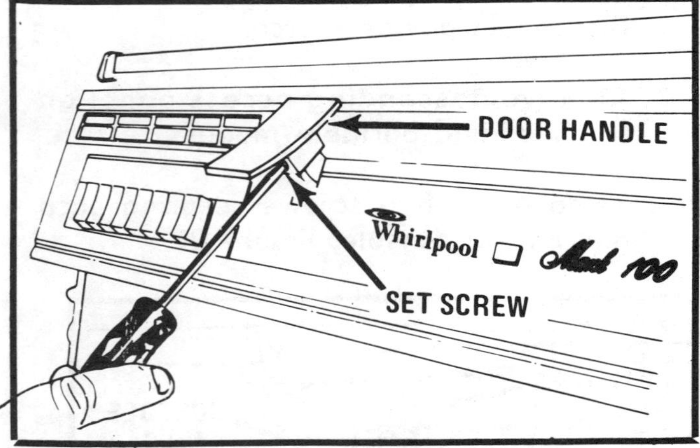

Figure 29

7. Remove two screws and rubber bumpers securing top of console to interior door panel, Figure 30.

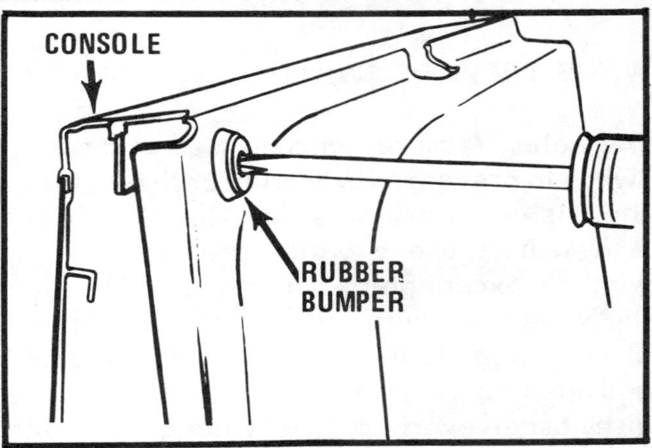

Figure 30

8. Remove timer knob and dial skirt from the timer shaft. (Not applicable on units with rapid advance timer system). Turn timer knob counter clockwise to remove.

9. Remove access panel assembly.

10. Remove wiring harness connections in the motor compartment and the door assembly. On rapid advance models, remove all the wiring harness connections from components in the door and console assembly.

11. Remove the screw that secures the wire harness to the lower door frame, pull harness up through opening provided in door frame, Figure 31. Models.

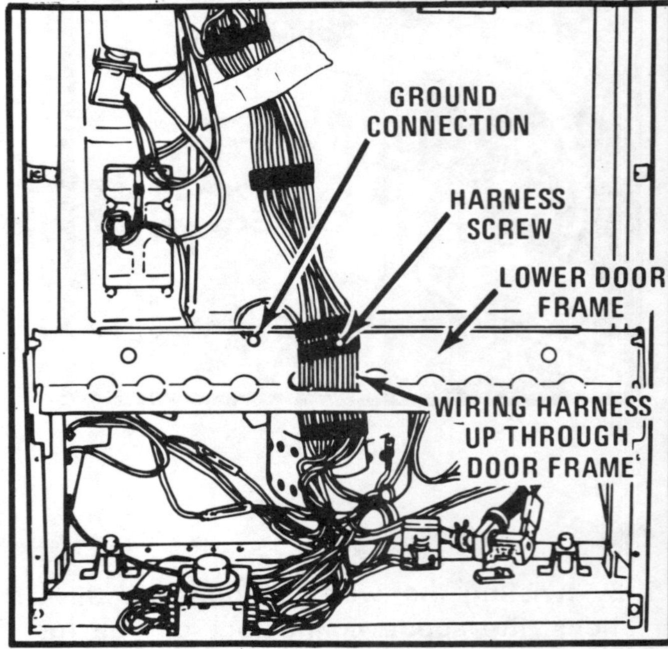

Figure 31

with rapid advance timer, pull harness down through opening provided in door frame, Figure 32.

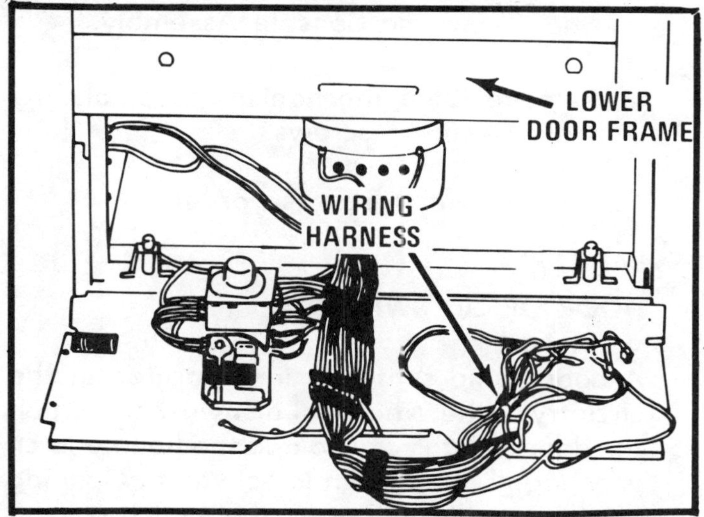

Figure 32

12. Remove two screws securing console to door frame, Figure 33, and lift console away from unit.

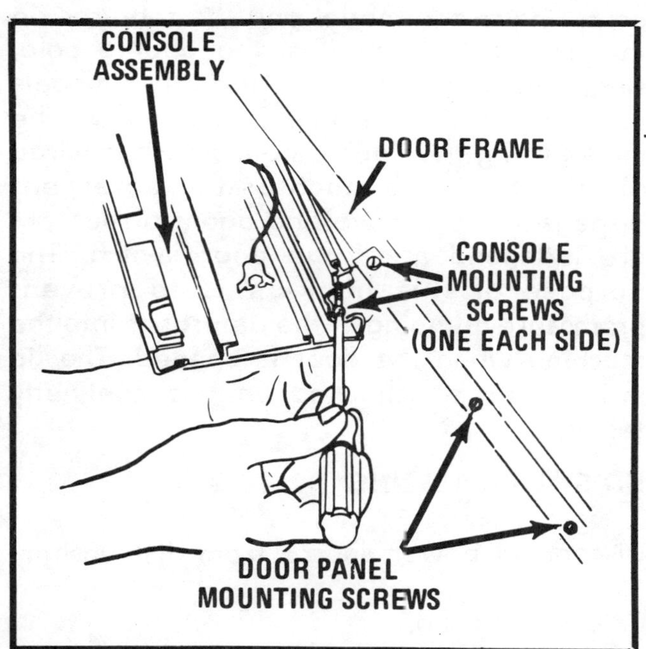

Figure 33

13. Release tension on door counterbalance springs, Figure 34.

14. Open door, remove four screws that hold the interior door panel to the door frame. Lift panel away from unit, Figure 33.

15. Remove counterbalance springs from extension of door frame, Figure 34.

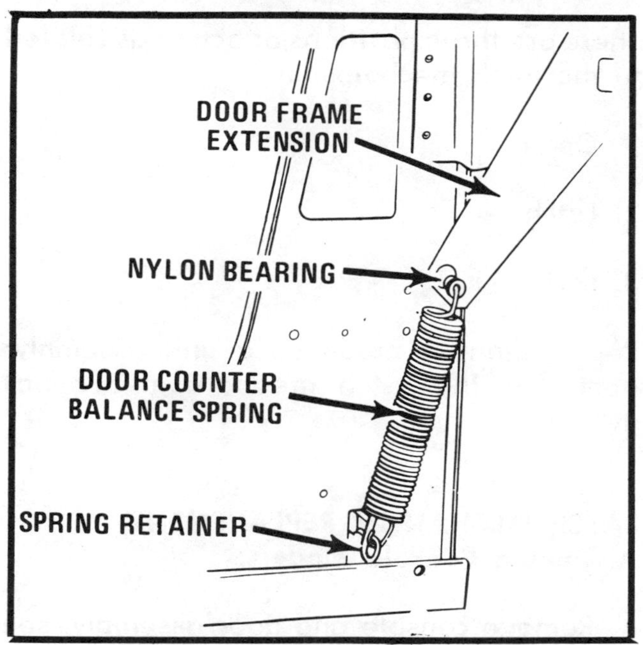

Figure 34

16. Partially close the door frame and lift up and out to remove it from the unit, Figure 35.

17. Reassemble in reverse order.

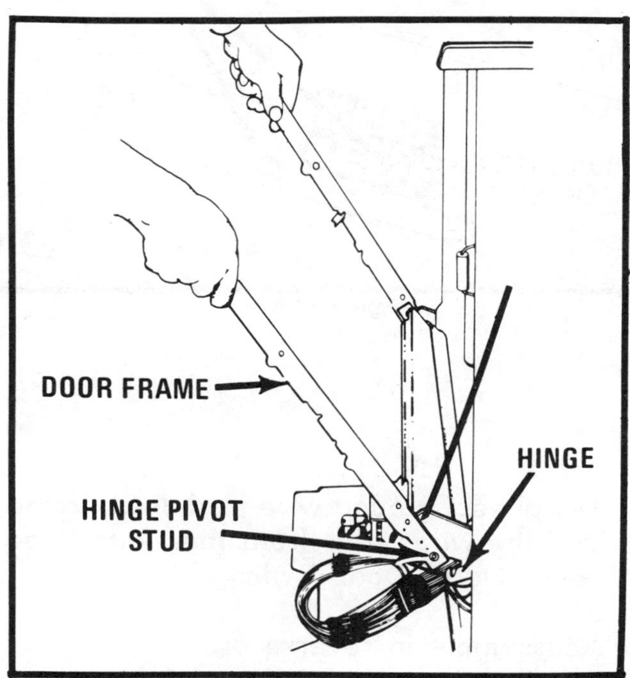

Figure 35

LATCH MECHANISM

There are three positions of action as related to the latch mechanism.

1. Open

2. Latch

3. Lock

Replacement is made as a unit assembly; parts for the latch mechanism are not available.

LATCH MECHANISM REPLACEMENT,
Aluminum Console Models.

1. Remove console and door assembly, see text, "Door and Console Assembly."

2. Remove the three screws that hold the latch mechanism to the interior door panel, Figure 36.

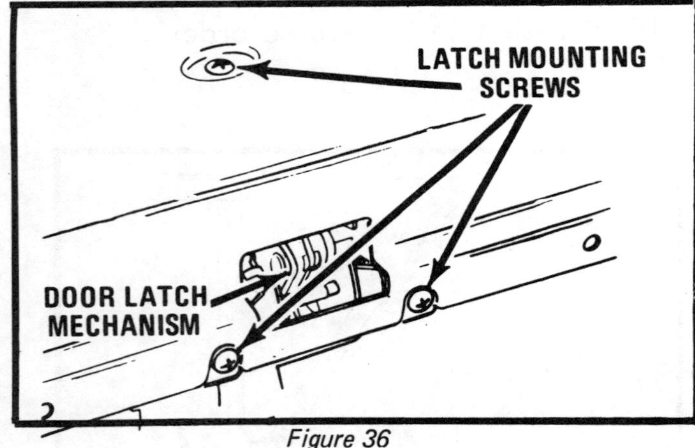

Figure 36

3. Disconnect ground wire from latch frame and the wire leads from the switch, see text, "Lid or Door Switch."

4. Reassemble in reverse order.

LATCH MECHANISM REPLACEMENT,
Plastic Console Models.

1. See "Door and Console Assembly."

2. Remove latch mechanism assembly, remove securing screws.

3. Reassemble in reverse order.

DOOR OR LID SWITCH

A door or lid switch is incorporated in the circuitry of the Whirlpool dishwashers. When the door or lid is closed and the handle latch is engaged, the switch is activated. Opening the door or lid will open the switch contact points, thus shutting down the dishwasher. The switch is a small micro-switch and is mounted on the door latch strike. The door handle will actuate the switch in the closed position. Two different switches are used according to the model and the features on the dishwasher; one uses a double pole, single throw switch, while the other models use a single pole, single throw switch. The model using the double pole switch is wired differently, in as much as the detergent dispenser and the wetting agent dispensers are independent of the door switch. The purpose of this hook-up is to prevent premature dumping of the detergent into the machine when the door is opened. The lid switch can be checked with a continuity tester.

LID SWITCH TESTING

Disconnect power source from dishwasher.

1. Open the lid.

2. Remove two screws that hold the strike and switch bracket.

3. Remove the two screws securing the switch to the switch bracket and remove the wire leads.

4. With a continuity meter or tester, place one probe on C for common, the other probe on N/O for normally open. There should be NO continuity. Depress button. With button depressed, the meter or the tester should show continuity. Remove probe from N/O and place on N/C. There should be continuity. Depress button. With button depressed, continuity should not exist. If the switch you are testing has only two terminals, select and use only the test that will apply, Figure 37.

NOTE: The leads of the switch are contained in a metal sheath below the tub gasket; the terminals or spade connectors are covered with a plastic sleeve, See Figure 37.

5. If a new switch is necessary, assemble new switch in reverse order starting at line 3, up to and thru line 1.

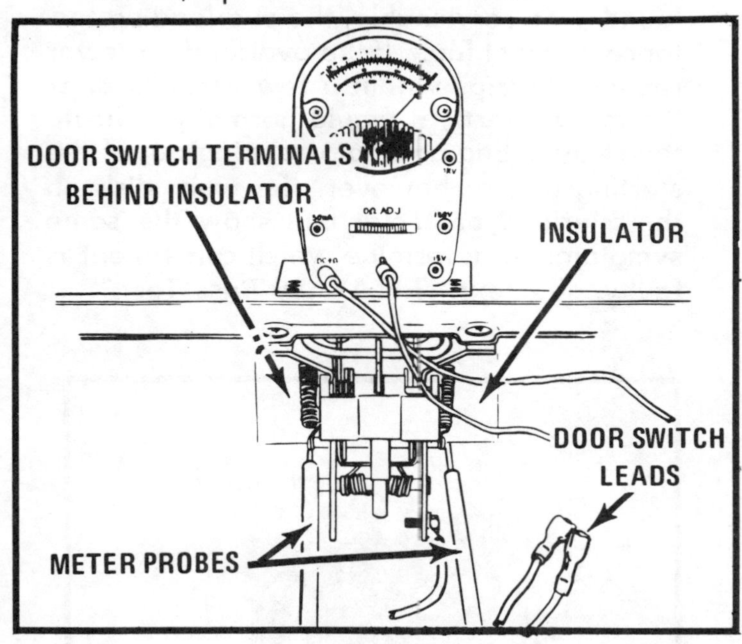

Figure 37

STARTING RELAY

The starting relay is a magnetic type of relay. The coil is in series with the motor "run" winding. The relay switch is in series with the starting windings, and the switch opens when the motor reaches running speed. The starting phase is then disconnected from the motor during dishwasher operation. Pick-up of the relay is at 12.8 amps, and drop-out

takes place at 10.5 amps. An inoperative relay, one that does not close the points of the starting winding switch, will cause the motor to "hang up" and a humming sound will be heard. This will generally trip the overload. The simplest way to check the relay is to by-pass the relay momentarily. See the following text, "Relay By Pass To Test."

RELAY, By Pass to Test, Figure 38

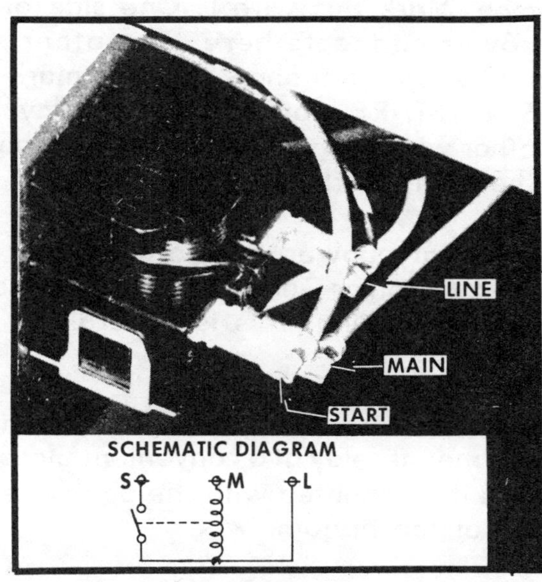

Figure 38

Remove wire from terminal marked "S" with timer in wash position. Touch wire from "S" to terminal "L". If motor starts immediately, remove the "S" wire, motor should continue to run if the problem was the relay. If motor fails to start or run, proceed to the following text.

REPLACING CENTRIFUGAL SWITCH WITH A RELAY, Early Models.

Follow procedures "Motor (Removing and Testing the Motor)."

a. Remove centrifugal switch from lower bell housing.

b. Remove wires from centrifugal switch.

c. Connect long leads to existing wires, about 10 inch leads, and mark them for identification. Reassemble motor.

d. With an ohmmeter, test across the leads.

e. From one of the three to another, look for a reading of 2.5 ohms. When these are identified, one wire will be the R for run or M for motor and the other wire of the 2.5 ohms will be the C for common or L for line. To determine which wire is L and which wire is M, find the pair that will read 5 ohms. The wire that is used to make the 2.5 reading and is also used to make the 5 ohms reading is the C or common. Mark this wire L. One side of the power connects here. The other side makes up the 5 ohms reading; mark that S for START marked on the relay. The other wire that makes up the 2.5 ohms is the RUN; mark this M.

Note the horsepower rating of the motor and secure a Klixon or Gemline relay of the same horsepower rating. This is generally a 1/3 horsepower relay.

f. Connect wires as previously identified and mount relay at a convenient place. Be sure it is mounted with the double terminal at the bottom.

g. Connect the line wires, one on the relay marked L, the other to the wire from the motor identified as common, and mark L.

The centrifugal switch, or the existing relay for the dishwasher motor, can now be replaced with the new solid state Gemline IC-21 relay. This relay is available at your local appliance parts supplier.

OVERLOAD PROTECTOR

All Whirlpool dishwashers have an overload protector in the electrical circuit. The protector is designed to open if excessive current passes through it. A manual reset overload protector is used on the impeller type of dishwasher. It is necessary to wait a few minutes before the reset button is depressed. This allows the warp to cool much the same as the reset in a service box. The reset will control operation of the motor and the inlet water solenoid.

All spray type dishwashers have an automatic overload protector built internally in the motor, Figure 39. When the overload protector in this type of dishwasher becomes inoperative, the motor must be replaced. On the impeller type dishwasher, if the motor and the relay have been tested and they are found to be serviceable, the overload protector could be at fault. If the overload protector repeatedly trips within a few seconds after the motor starts, it would normally indicate the relay is bad and not dropping out of the starting phase; however, the warp disk on the overload protector can show the same symptom. To determine which component is faulty, see text, "RELAY, By Pass Test."

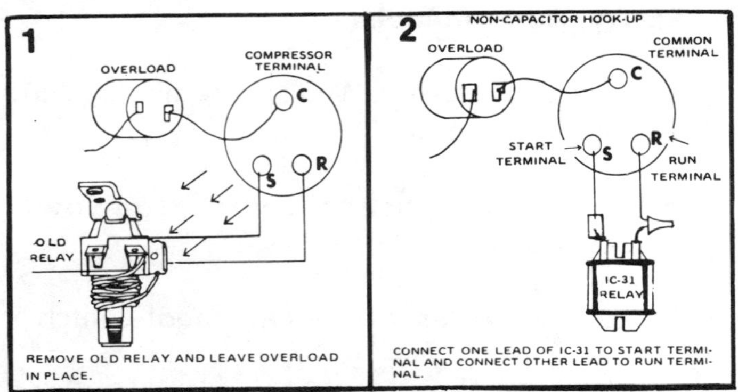

Gemline IC- Relay

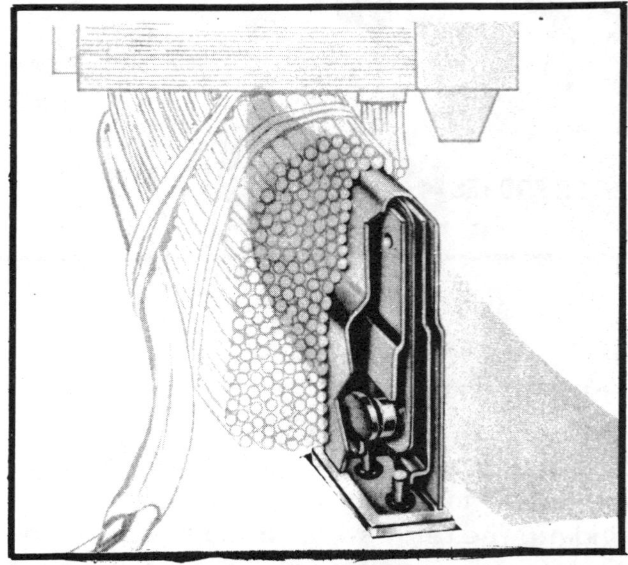

Figure 39

SOLENOIDS

A coil of wire is wound spool like, and insulated with a hole through the center, not unlike a spool of thread. This coil in early models was installed in a metal case, Figure 40. In later models the coil was encapsulated

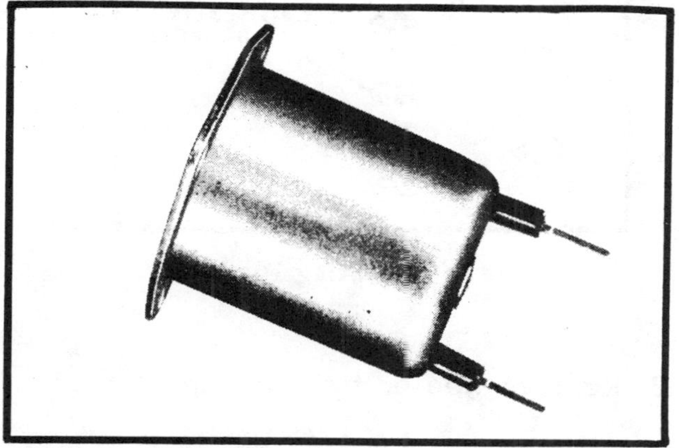

Figure 40

in an epoxy or resin like insulater, Figure 41B. In still earlier models the coil was insulated with a cotton tape and varnished, Figure 41A. The principle of the solenoid coil was also used on early models to activate the drain valve on the models that did not use a drain pump, see "SOLENOID TEXT."

This coil as used on the water valve is placed over a brass cylinder that contains a plunger, often referred to as an armature. One end of the plunger is tapered and used as a needle valve. A push spring is placed at the other end of the plunger. When the coil is energized the magnetic force lifts the plunger from the seat allowing water to enter the valve through the orifice. Because of the bleeder hole in the diaphragm, the diaphragm is lifted from the seat and hangs in limbo, allowing a full flow of water to enter the tub.

If a condition exists that water enters the tub in other cycles than intended, there is an indication that the valve is mechanically stuck open and must be disassembled and cleaned.

CHECK POWER TO THE WATER VALVE

Test with power on, set the timer to the FILL cycle, and place a test lamp across the terminals of the solenoid coil. If the lamp lights, it indicates power is reaching the coil, and the timer is not at fault, nor has the water level switch tripped. Lack of water flow would be due to an internal problem of the water valve, or inoperative solenoid coil, Figure 42.

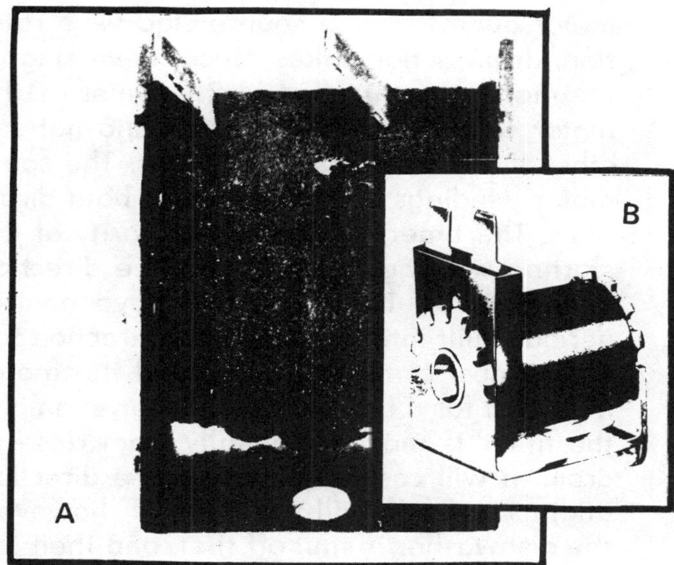

Figure 41AB

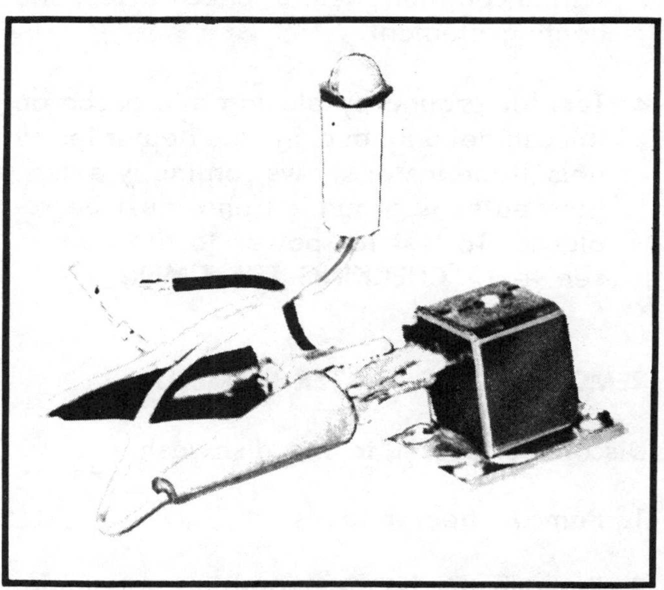

Figure 42

By removing one wire from the solenoid coil and placing the test lamp leads between the solenoid terminal and the wire that was removed, the coil can be tested. If the lamp glows slowly and dull, the coil is serviceable; if no light shows in the lamp, the coil is open and must be replaced.

HEATERS

All Whirlpool dishwashers have a cal rod type heater assembly that is mounted to the bottom of the tub with gaskets and brass nuts. The heating element maintains the temperature of the water during operation of the dishwasher, and also heats the air for drying the dishes when the "DRY" cycle is used.

The heating element on the impeller type of dishwasher is rated at 600 watts; the spray arm dishwashers use a heating element rated at 800 watts. Both types of elements mount the same way.

TESTING THE HEATER

1. Remove the power to the dishwasher.

2. Remove the wire leads from the heater.

3. With a continuity tester, check across the heating element.

4. Test for ground by placing one probe on the cabinet and touching the heater terminals. If the tester shows continuity exists, the heater is grounded and must be replaced. To test for power to the heater, see text, "CHECKING THE TIMER."

REMOVING THE HEATER ELEMENT

Disconnect power to the dishwasher.

1. Remove heater leads.

2. Remove the heater mounting nuts under the tub, Figure 43.

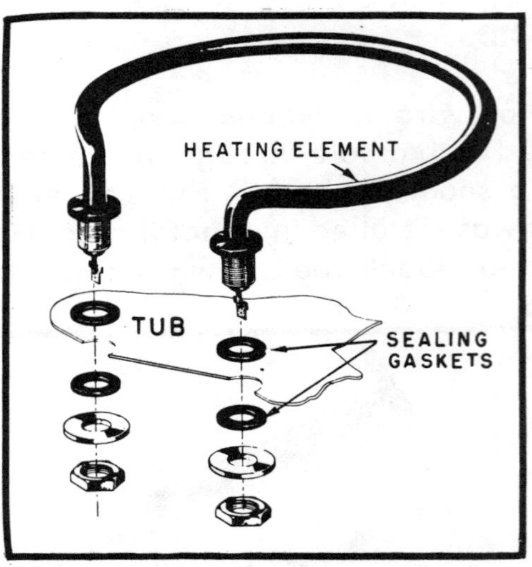

Figure 43   Heating Element

3. Remove heater from inside of tub.

4. Replace in reverse order.

MAIN DRIVE MOTOR

The impeller type dishwashers are equipped with a 1/3 H.P. 1725 R.P.M., reversible motor. A remotely mounted overload protector with a manual reset protects the motor from excessive overloading. When the motor is rotating in a clockwise direction, washing and rinsing action takes place. When the motor operates in a counterclockwise rotation, drain action takes place. Reversing of the motor is controlled by the timer. If the motor rotates in one direction and not the other, the fault is NOT the motor. The same motor windings are energized in both directions. The timer changes the polarity of the starting winding; this changes the direction of the motor. The timer will have neutral periods built into it. To reverse direction the motor must come to a stop, and the motor must rest for a few seconds. If, for example, the timer is moved manually from rinse to drain, it will continue in the same direction and, as a result, will not drain. If, however, the dishwasher is shut off first, and then the timer is moved to drain, it will turn in the opposite direction and drain when started.

TESTING THE MOTOR

Disconnect power to the dishwasher.

1. Remove the blue motor lead from the relay.

2. Disconnect the motor leads at the motor; this will be a yellow, red and black wire.

3. Connect the red and yellow wires together temporarily.

4. Using a direct line test cord, connect one side of the line to the red and yellow wires and the other side of the test cord to the blue wire.

5. Touch the black wire to the blue wire as an assistant connects the test line. When and if the motor starts, immediately remove the black wire from the blue wire. Take note of the direction the motor is rotating.

6. To reverse the direction, move the black wire to the yellow wire and momentarily touch the red wire to the blue wire as an assistant connects the test cord. Immediately after the motor starts, remove the red wire from the blue wire.

7. If motor hums and fails to run, check the relay.

The motor can also be tested with a three wire test cord such as the Gemline TC-6, which is available at your appliance parts supplier. To use the Gemline test cord to test the motor, proceed as follows:

TEST CORD LEADS		MOTOR LEADS
Black	To	Yellow and Red
Red	To	Blue
White	To	Black

Connect cord, depress button momentarily; if motor operates take note of direction.

To reverse direction change the positions of the Red and Black motor leads and repeat the test.

The motor should be tested for ground with the use of a series test light. Connect one probe to the motor casing and touch each motor lead separately. If the lamp lights on any part of this test, the motor is grounded (a wire in the motor touching the casing) and should be replaced.

If you are testing the motor while it is still in the dishwasher, be sure the pump is clear and not binding the motor. Spin the impeller manually before making any test on the motor. If the motor is binding, it must be cleared before any test can be made.

MOTOR REMOVAL, Horizontal Motor.

1. Remove rear cabinet shroud.

2. Check power to the motor before removal.

3. Disconnect power. Using a padded surface lay dishwasher on its side.

4. Disconnect electrical leads from motor, and inlet and drain hoses from pump.

5. Remove the four main motor mounting screws that secure the motor to the base, Figure 44.

6. Lift motor free of the machine.

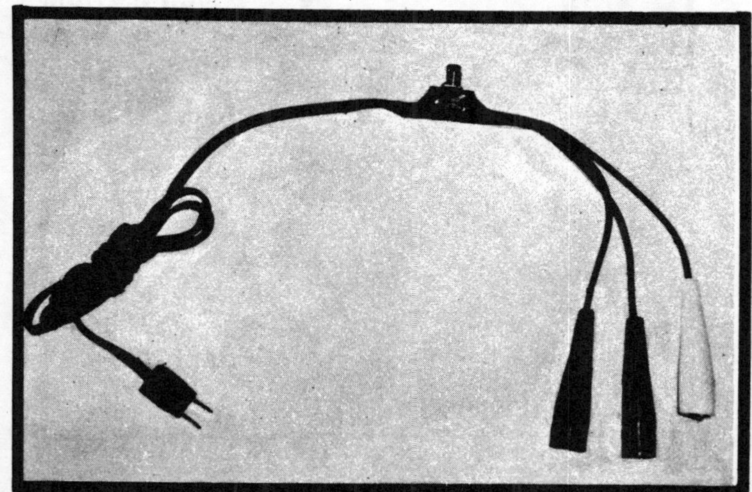

TC-6 Six feet heavy duty three-wire test cord with pushbutton start switch 20 amp. capacity.

Figure 44

MOTOR REMOVAL, Vertical Motor.

1. Remove read cabinet shroud, remove impeller inside tub.

2. Check for power to the motor before removal.

3. Disconnect power. Using a padded surface lay dishwasher on its side.

4. Remove electrical leads and hoses from pump.

5. Support motor, loosen motor clamp mounting screws, Figure 45, remove mounting clamp, Figure 45.

6. Remove motor from cabinet. To install, reverse procedure. Care should be taken that the impeller shaft is centered in the sump hole.

PUMP MOTORS, Late and Current Models.

The pump motor on the late and current models of Whirlpool dishwashers are quite different in design and operational characteristics than early models. The pump motor is a 1/3 H.P. and rotates at 3450 RPM. It is rated at 120 Volts A.C., 60 Hz. The rated load is 6.0 amps, 570 watts. A magnetic relay is used in the starting phase, and the automatic protector is built into the motor windings. This is a two directional motor, and control of the direction is through the timer. It has two separate phase windings and rotates clockwise for water circulation and counterclockwise for draining.

PUMP MOTOR TESTING

Test equipment, as shown below, is available through your local parts distributor, Figure 46. Order Robinair Part No. 14345; however, the motor can be checked as follows

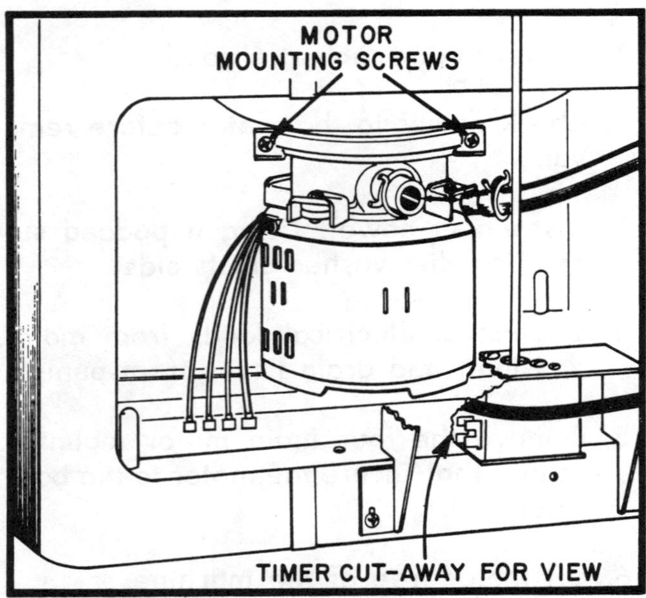

Figure 45

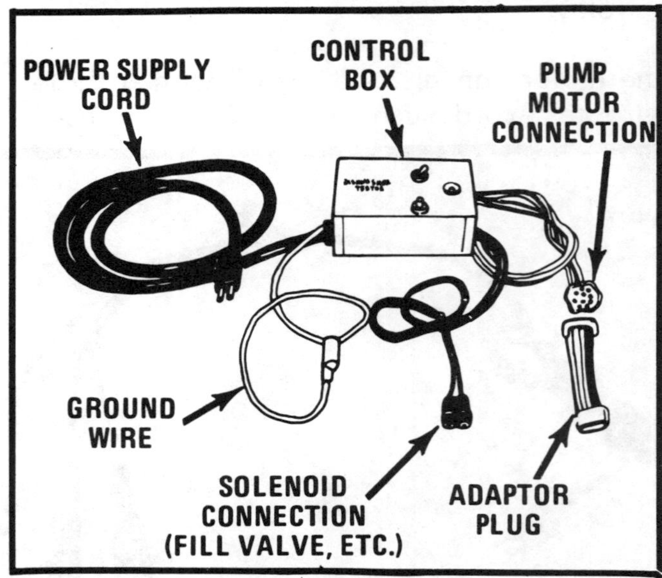

Figure 46

1. If motor fails to run, check the following.

2. Check all controls for operating position.

3. Check indicator lights, water fill, etc; if nothing operates, check the fuse or breaker switch.

4. If motor hums, but doesn't run, bypass the relay, see text, "Relay Bypass."

5. If everything but the motor operates, disconnect power source.

6. With an ohmmeter check the motor windings for continuity; also check for grounded motor.

7. If motor runs but repeatedly cuts out, check pump for jammed impeller or foreign matter in the pump.

8. If motor does not reverse, check timer contacts. On the timer "Y" is the clockwise starting winding connection, and "GY" is the counterclockwise starting connection.

9. Check timer for proper switching.

PUMP MOTOR, OHMMETER TESTING

Set Ohmmeter on RX1 scale for continuity test.

1. Place one probe on terminal #4 and the other probe on terminal #1; reading should be 2.7 ohms.

2. Place one probe on terminal #4 and the other on terminal #2; the reading should be 5.6 ohms.

3. Place one probe on terminal #4, the other probe on terminal #3; the reading should be 5.6 ohms.

4. Place one probe on the motor case (an area free of paint), the other probe place momentarily on each terminal. If continuity shows, the motor is grounded and must be replaced, Figure 47.

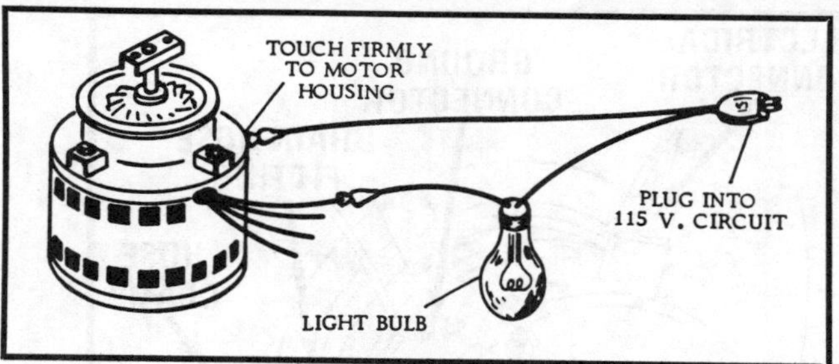

Figure 47

5. In the above tests an open winding or an open protective device will not register resistance on the ohmmeter; an over or under reading will indicate shorted windings. In either case the motor will have to be replaced.

PUMP MOTOR, Replacement.

1. Disconnect power source.

2. Remove dishracks, spray arm, and filter screen from dishwasher.

3. Remove access panel toe plate assembly from undercounter models. Portable can be placed on its back to gain access to the pump motor.

4. Remove electrical connector, ground wire and drain hose from pump assembly. Make preparation to catch water that's in the hose, Figure 48.

5. Remove drain hose fitting from pump mounting plate.

6. Remove four pump mounting clips and lift pump out.

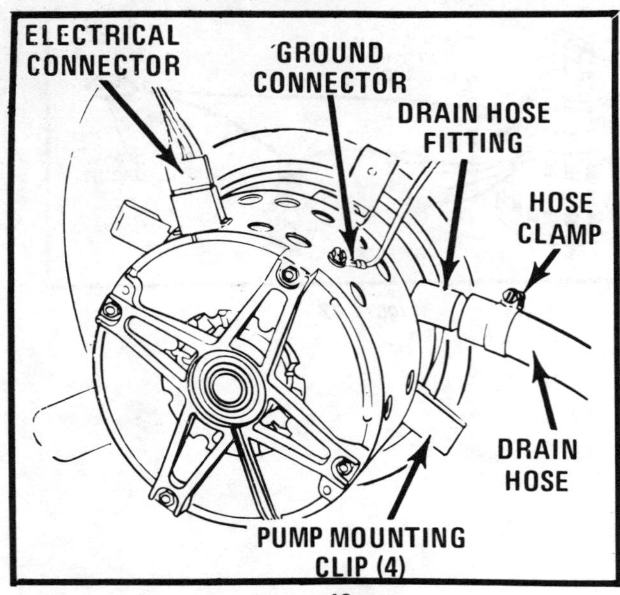

Figure 48

DRAIN PUMPS, Testing, Repairing or Replacement.

Removal, Figure 49.

a. Disconnect wiring from motor.

b. Remove four metal screws holding pump motor to bracket.

c. Remove pump.

Testing

a. Remove pump cover screws, Figure 49.

Figure 49 - Pump and Motor - Exploded View

b. Remove pump cover.

c. Insert small screw driver, Figure 50, between rotor and the field laminations to keep rotor from turning.

d. Unscrew small impeller.

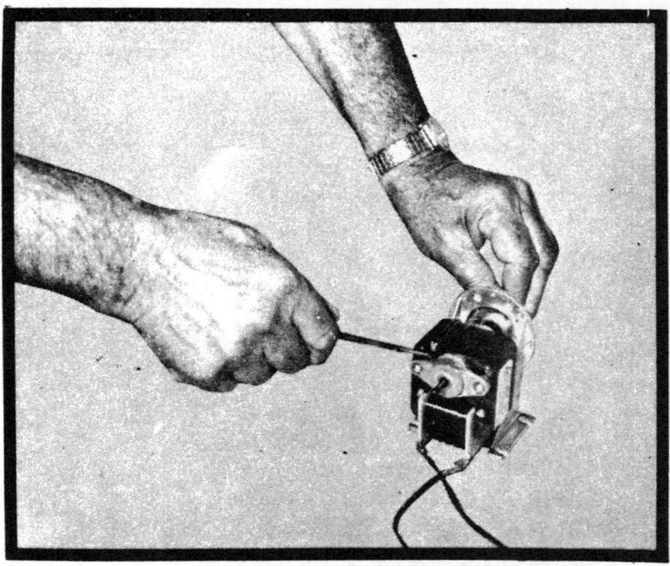

Figure 50

e. Remove diaphragm seal, being very careful not to tear.

f. With a fine grade of oil, lube both end bearing plates.

g. Note if shaft turns freely by hand.

h. Connect cord directly to motor and test motor.

i. If line (g) and (h) test good, the motor is still serviceable.

j. Check bearings for wear.

k. If pump motor is good and a condition exists that the seal leaks around the shaft, proceed as follows:

PUMP DIAPHRAGM SEAL REPAIR

If a new diaphragm is not available and the shaft seal leak is the problem, it can be repaired. If the diaphragm is torn, it is best to replace the component part, available at your appliance parts distributors.

a. Clean and polish shaft where lip seal rides.

b. Clean and wipe dry diaphragm; place it over shaft and align to groove in motor frame.

c. Carefully wipe some Pliobond or rubber cement to the outer edge of lip seal, Figure 51.

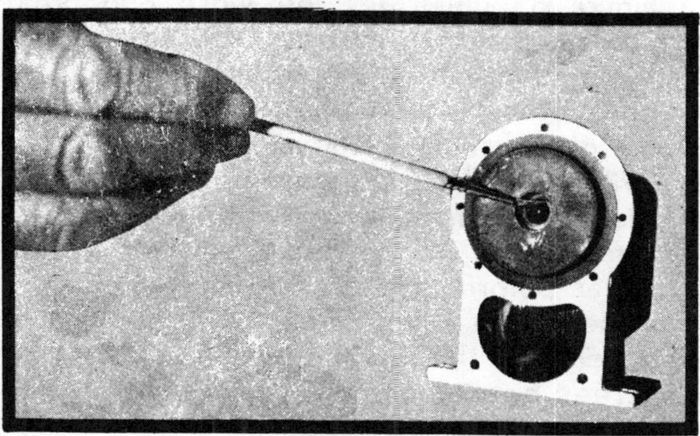

Figure 51

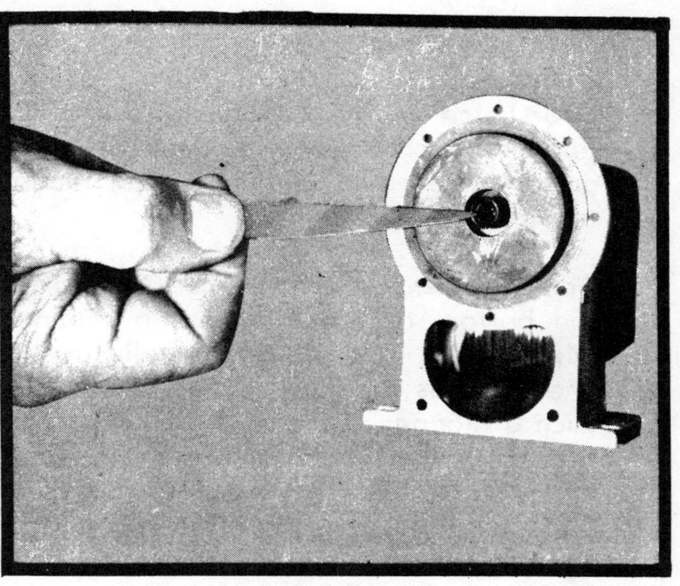

Figure 52

d. Place a 1/4" diameter o-ring over extended edge of lip seal. (If you have the heavy shaft pump motor, a larger o-ring will have to be used), Figure 52.

e. Take note that the o-ring has tightened the shaft hole to prevent a leak.

f. Replace small impeller; note the impeller will hold the o-ring in place.

g. Reassemble pump cover to base.

h. Plug center outlet with a cork. Fill pump with water and test for leaks.

i. Reassemble to machine; connect wiring and hose.

UNDERCOUNTER MODELS

1. Door Latch — The handle, latch and interlock switch are mounted on the door and engage an adjustable striker mounted on the tub.

2. Door Gasket is in two parts. One across the bottom of the door has a retainer with slotted holes to allow adjustment. The other is mounted inside the tub in such a manner that the top and sides of the door will seal against it to prevent water leakage.

3. Door Hinges are mounted at each of the bottom corners of the door and are spring loaded to counterbalance the door. They should be so adjusted as to cause the door to want to closed.

PORTABLE

1. Lid Latch — The handle and lid latch mounted on the lid engage the striker and interlock switch mounted on the cabinet. Adjustment is made on the striker.

2. Lid and tub gaskets provide a seal to prevent excessive water leakage. Both are readily replaceable.

3. Lid Hinges include an adjustable counterbalancing spring.

4. Detergent Dispensers — Model 40 and 50 series are mechanically controlled by a cam on the timer shaft. As the timer advances, the cam allows a plunger to retract, dumping one cup in the first wash cycle and the other in the second wash cycle.

5. Casters — all portable models ride on casters. Casters are bolted to the cabinet and may be readily removed for servicing.

6. Access panel for these models is on the back of the machine, held in place by four screws.

WATER CIRCULATION, Spray Arm Method.

Figure 53 illustrates the wash and drain system of the spray arm type dishwasher. The left side of the illustration shows dishes and glasses in their relative positions when the dishwasher is loaded. The water, after striking the dishes, drains into the bottom of the tub, which slopes toward the sump. The sump screen strains food particles from the water before it returns to the sump. The slope of the sump screen and the flow of water wash the food particles into the sump screen basket, which traps and contains large particles.

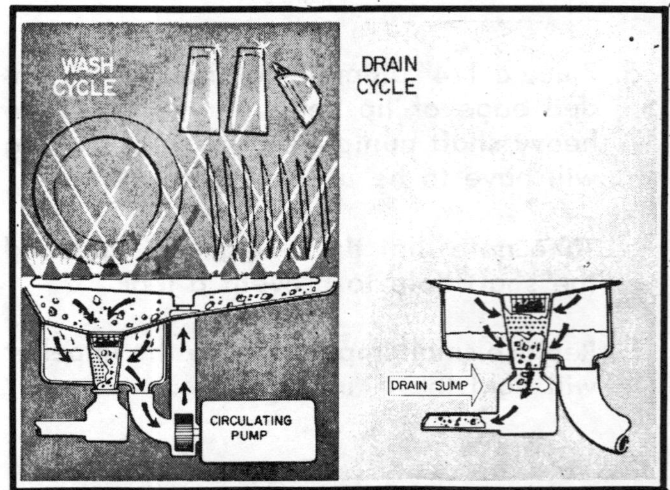

Figure 53

Small particles pass through the basket grid and into the cone-shaped portion of the screen, which empties into the drain sump. During pump-out, the drain pump withdraws the water from the sump through the cone-shaped screen, which washes it free of particles that might be in the screen mesh. This makes the cone-shaped portion of the screen self-cleaning; the circulating water during operation keeps the top portion clean. The sump screen basket should be withdrawn and emptied, if necessary, after the complete cycle is finished.

The filtered water in the sump is pumped upward into the spray arm and out the nozzles in the arm. The force with which the water is ejected from the canted nozzles causes the arm to rotate.

The centrifugal type circulating pump is mounted on the main motor and circulates 30 gallons of water per minute.

The drain pump, connected to the pump reservoir, is also a centrifugal type pump that is driven by an attached shaded pole motor.

Impeller Method

The impeller method of dishwashing employs a motor-driven impeller that scoops up water and throws it upward and outward. This strong water action washes and rinses the dishes. Figure 54 illustrates the wash and drain system of this type dishwasher. The impeller is driven directly by the motor and rotates at motor speed, 1750 R.P.M. Drain action takes place when the motor direction is reversed. When the motor is reversed, the drain pump (shown below the impeller in Figure 54) discharges the water from the

dishwasher. During the wash the impeller turns clockwise, and during the drain it turns counterclockwise.

VERTICALLY MOUNTED
Wash and Drain, Pump Motor Assembly

NOTE: Because of the similarity of mechanical design, the following text will apply to the repair of both the early (Figure 54A) and late (Figure 54B, 55) Delco design units.

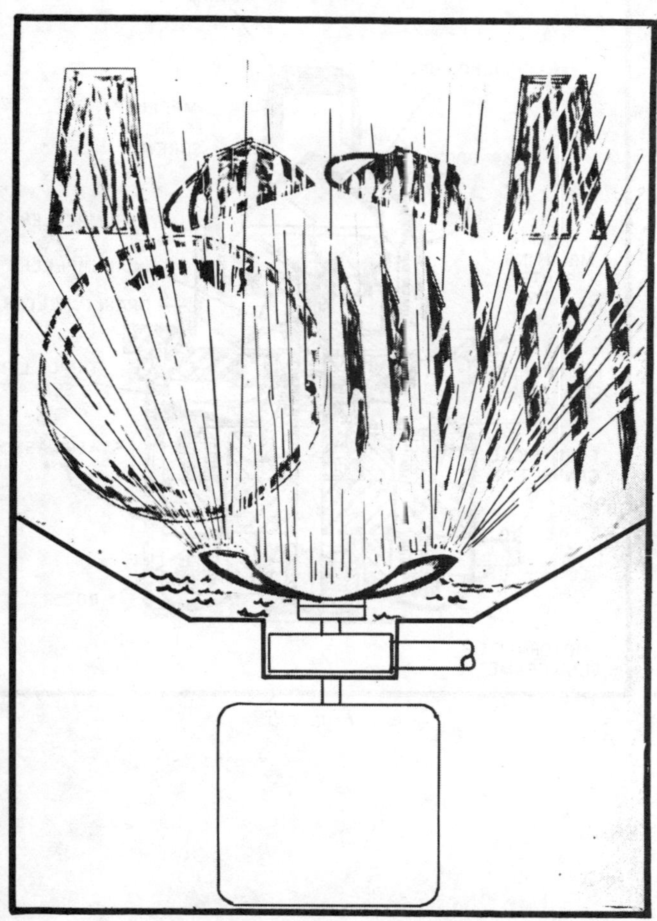

Figure 54

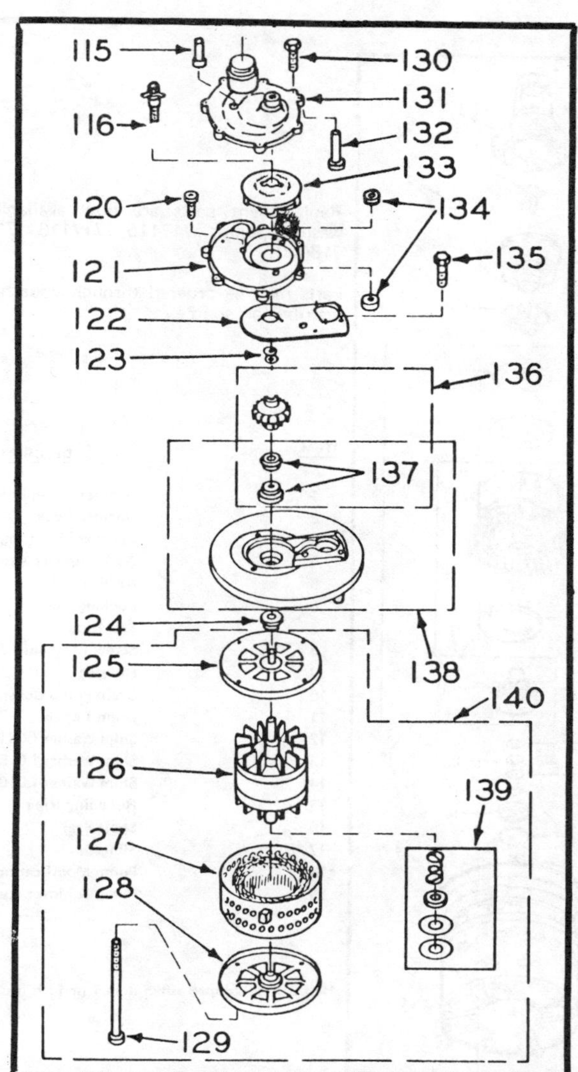

EARLY DELCO DESIGN UNIT

Figure 54 A

115		Pipe, Stand
116		Screw, Impeller
		10-32 x 2"
120		Screw, 8-32 x 1 1/8 (Package of 3)
121		Base, Volute
122		Cover, Drain Pump
123		Washer, Shim (Package of 4)
124		Spacer, Impeller
125		Frame, Pump Side

126	Rotor
127	Stator - 60 Cycle
	Stator - 230 Volt, 50 Cycle
128	Frame, Outer
129	Bolt, Thru (Package of 4)
130	Screw, 8-18 x 1" (Package of 7)
131	Cover, Volute
132	Tube, Vent
133	Impeller, Wash
134	Washer, Vent (Package of 2)
135	Screw, 8-32 x 1/2
136	Impeller
137	Seal
138	Plate, Pump Mounting
	Plate, Pump Mounting (Pump 713147)
	Plate, Pump Mounting (Pump 713105)
139	Thrust Spring and Washers
140	Motor

WATER CIRCULATION COMPONENTS
Circulation And Drain Pump Assembly.

1. Circulation can be checked by the sound created of water striking the tub at regular intervals as the spray arm rotates.

2. The rotation can be timed. The spray arm rotates at 25 R.P.M. plus or minus 5 revolutions.

3. The dishwasher should drain in 40 seconds or less. The drain lines should be checked and cleared before this test.

4. The efficiency of the drain cycle can also be checked by visual examination after a drain cycle. Open the door right after the drain period and remove the drain sump filter. Water line should not exceed ½ inch up on the plastic filter screen.

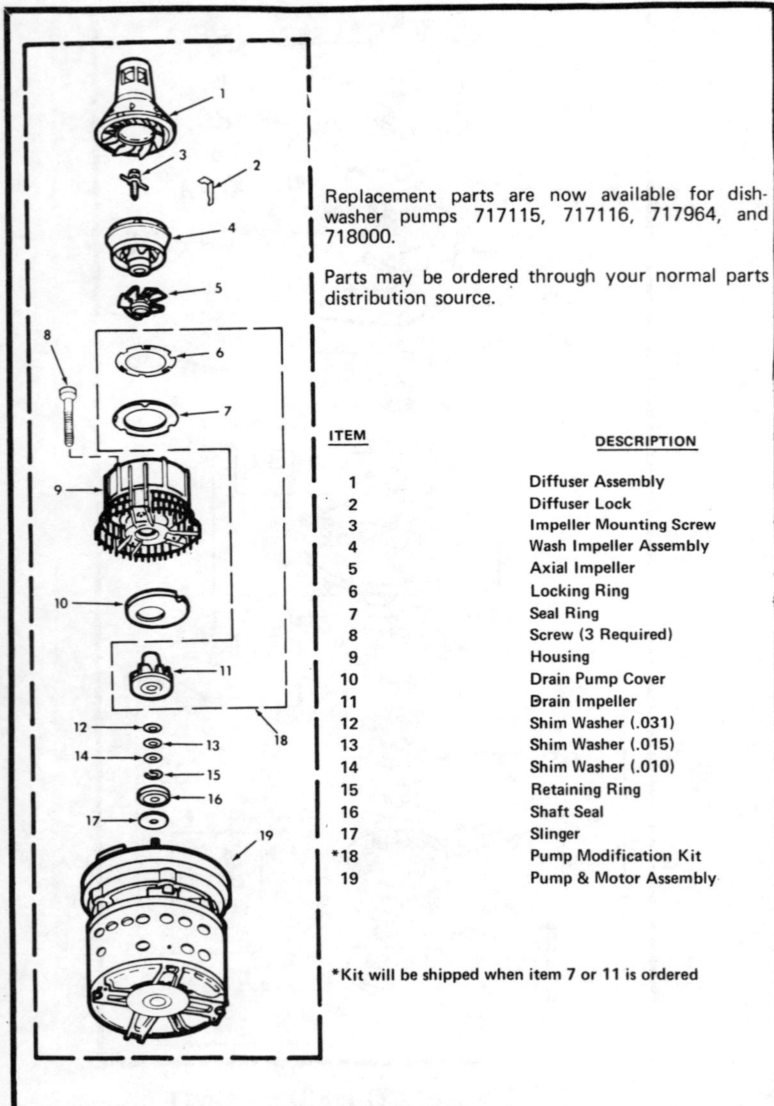

LATER DELCO DESIGN UNIT

Figure 54 B

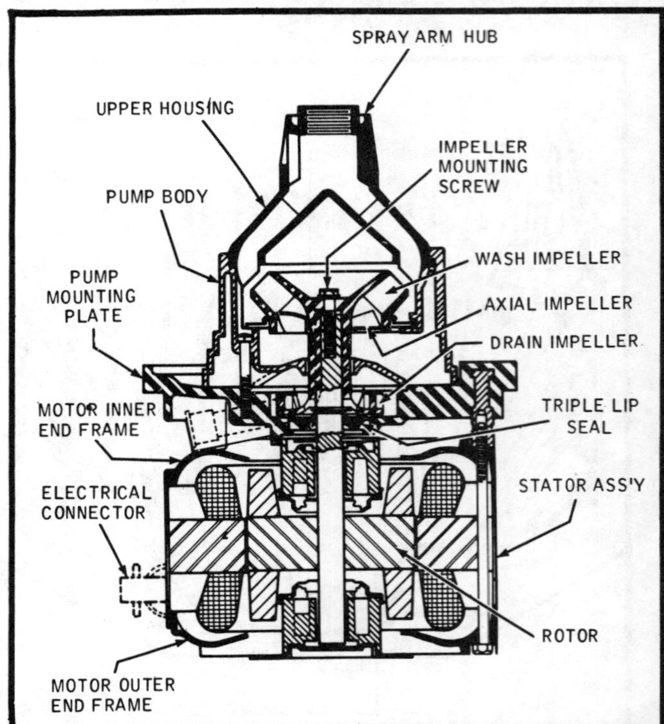

Figure 55

DRAIN PUMP, Replacement.

The pump parts are replaceable only. In the case of motor failure the complete motor and pump must be replaced. Do not remove the pump mounting plate from the motor stator assembly. The motor shaft seal tolerance must exceed .010 inch maximum in order for the lip seal to be effective. If the through bolts of the motor are disturbed, this tolerance could be lost and cause the seal to malfunction.

DISASSEMBLY

1. Remove diffuser lock from base of diffuser cover. Do not discard the diffuser lock, see Figure 56.

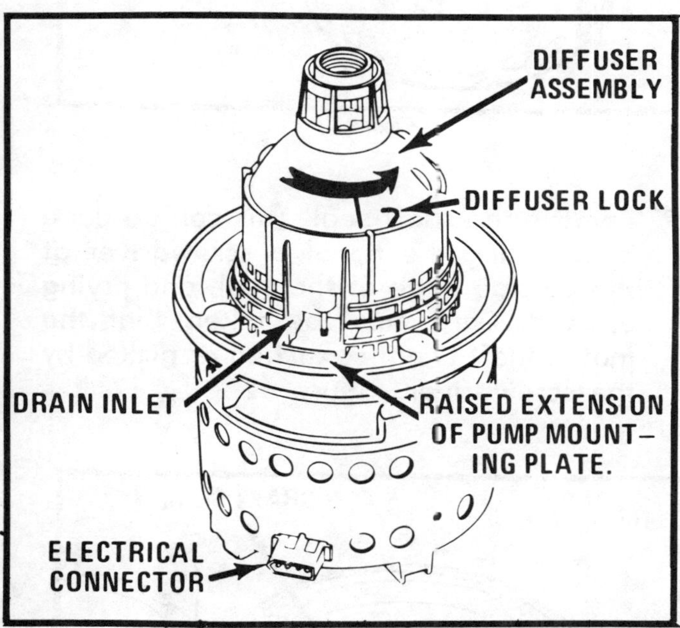

Figure 56

2. Rotate diffuser assembly counterclockwise and lift out of locating slots.

3. Bend locking tab up on impeller mounting screw and remove mounting screw. The screw has a right hand thread.

4. Remove the impeller off of the motor shaft, see Figure 57.

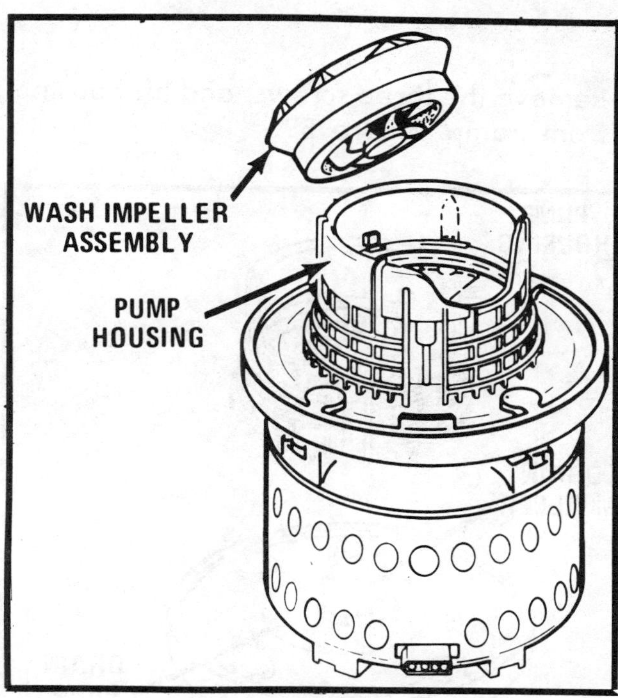

Figure 57

5. If the three pump housing screws are visible, it would indicate that the pump has the new style locking ring which will not require removal, Figure 58, however,

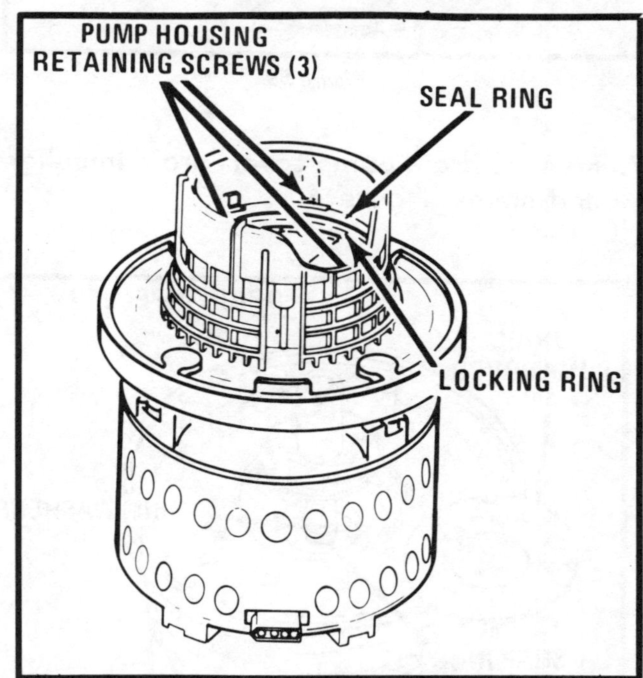

Figure 58

it the screw heads can not be seen, the lock ring must be removed to gain access to the screws. The locking ring will be destroyed upon removal. Replace with part number 718342.

6. Remove the three screws, and lift housing from pump, Figure 59.

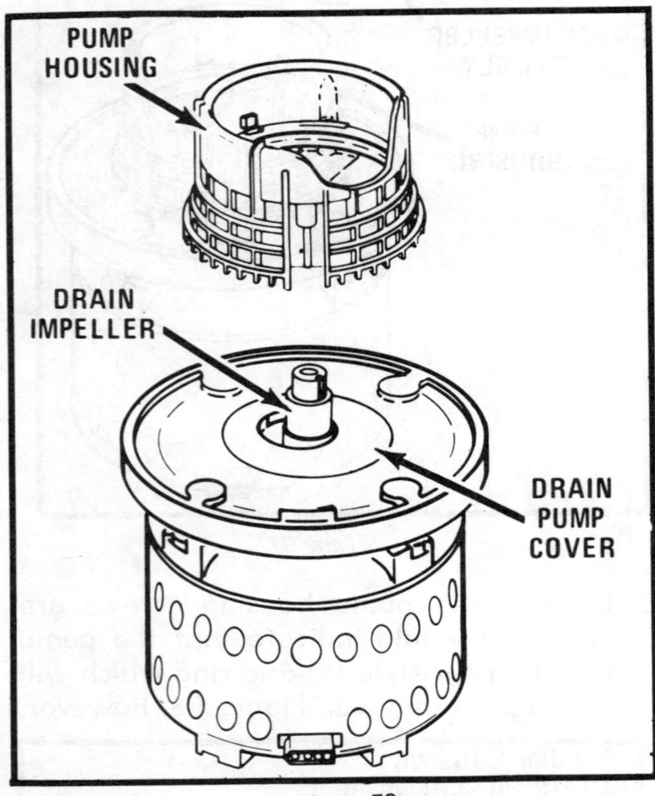

Figure 59

7. Remove drain pump cover, drain impeller and shims, Figure 60.

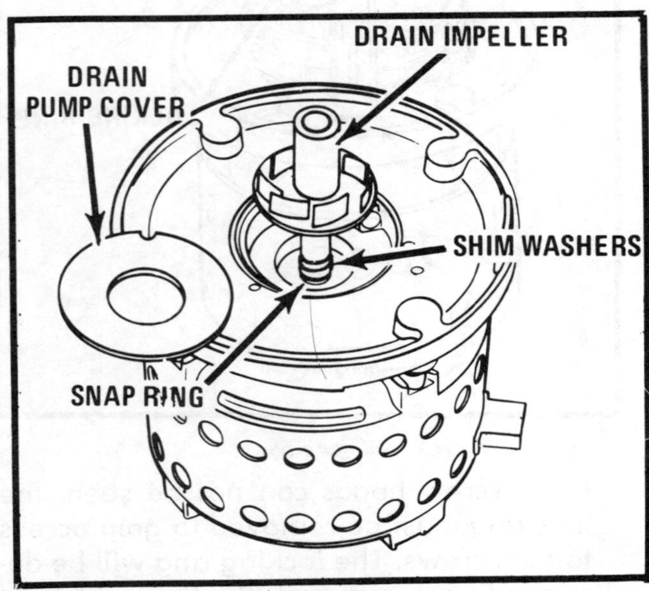

Figure 60

8. Remove retaining snap ring from motor shaft; this ring is to be replaced with part number 718339, Figure 61.

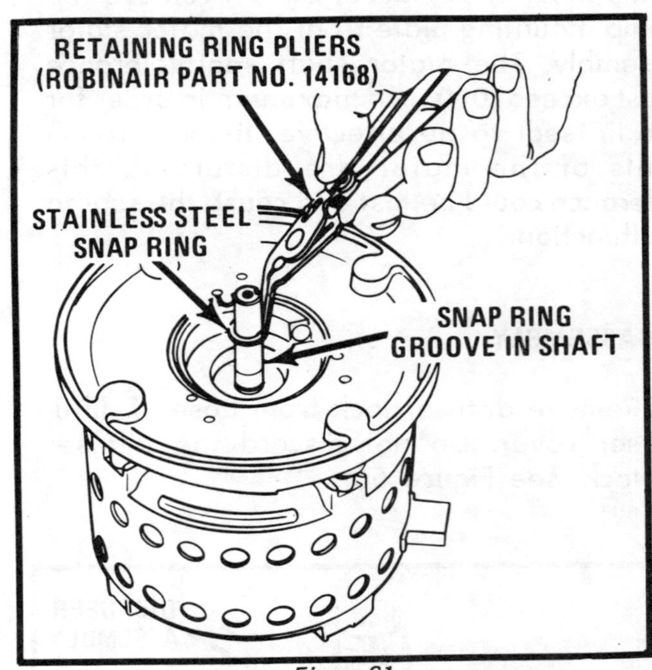

Figure 61

9. Remove the shaft seal. This can be done by inserting the tip of a screwdriver at the outside edge of the seal, and prying upward. Care must be taken that the motor shaft is not scratched or nicked by the screwdriver, Figure 62.

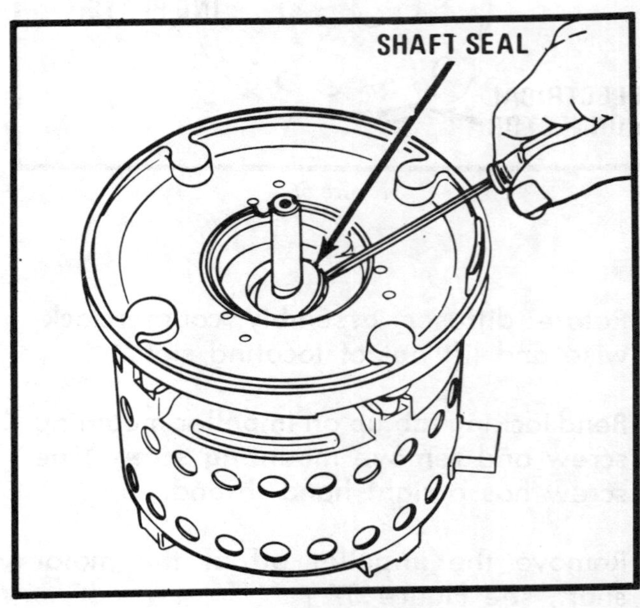

Figure 62

REASSEMBLING THE PUMP MOTOR

The following should be checked prior to reassembly.

1. End thrust of the motor shaft should not exceed .030 of an inch, Figure 63.

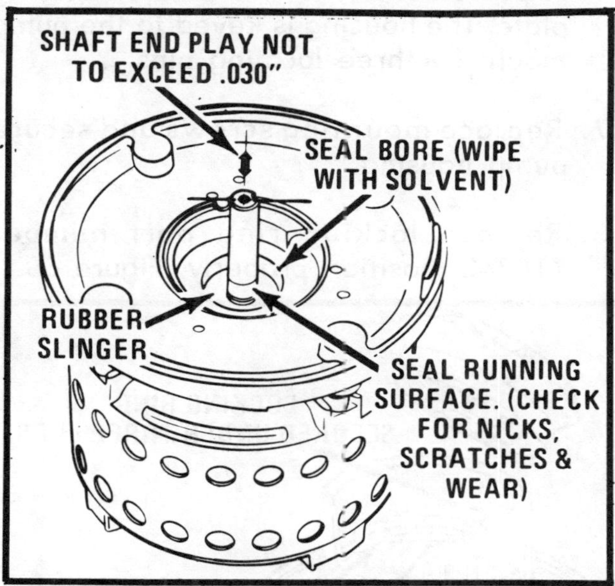

Figure 63

2. Free rotation of the shaft, check for bearing wear.

3. Test motor, see "MOTOR TEST" text.

Remove the rubber slinger from the motor shaft and add a few drops of oil to the bearing reservoir. Oil should be a good grade SAE No. 20, and non-detergent. Reinstall slinger ring in motor shaft groove.

4. Wipe the seal bore of the pump plate clean, using a solvent.

5. Check motor shaft where seal lip rides for scratches and nicks. If shaft is damaged or worn, the entire pump and motor assembly must be replaced, Figure 63.

6. Wipe outside edge of replacement seal with solvent and remove oil and grease, wipe dry.

7. Wipe a bead of pliobond around outside edge of the new seal, also inside the seal cavity; allow to set a few minutes before installing the seal into the seal cavity, Figure 64.

8. Insert seal protector, supplied with seal kit in motor shaft key-way before installing new seal. This prevents the sharp edge of the key-way from damaging the inner lips of the seal during installation, Figure 64.

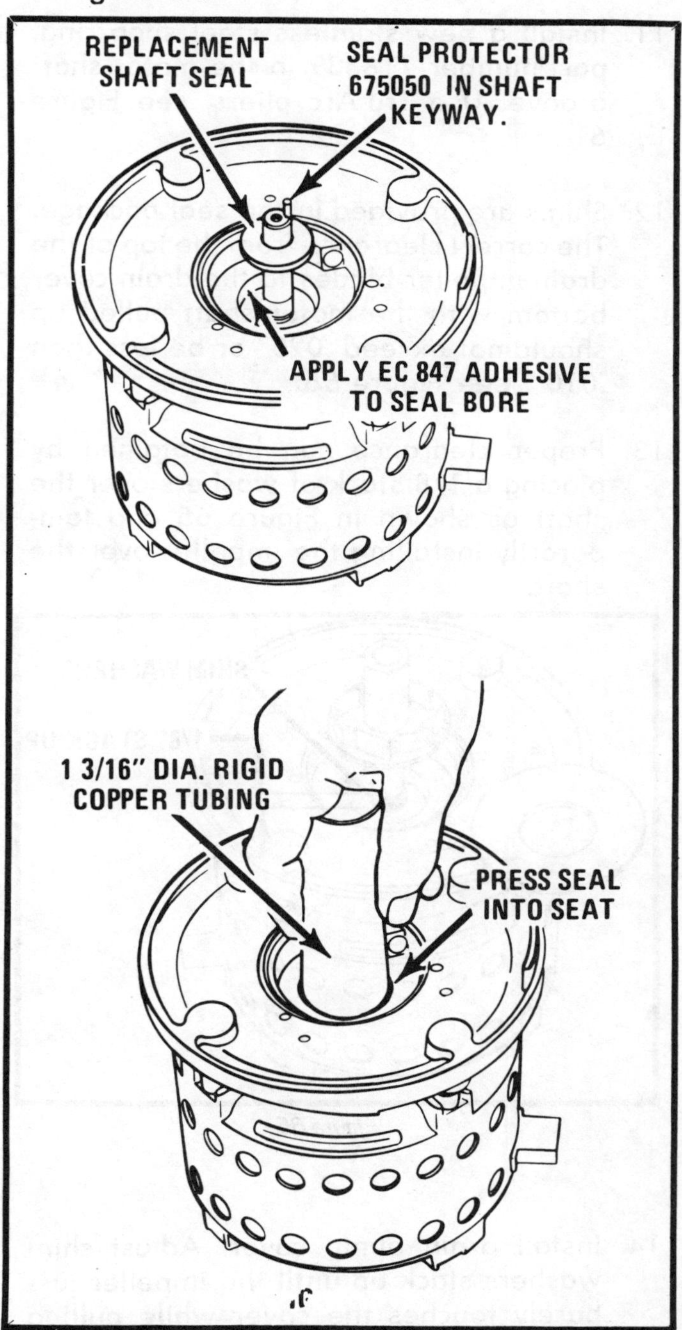

Figure 64

9. Install seal as marked "This side up;" be sure that the seal is pressed all the way down in the seal cavity. A 4 inch length of 1-3/16 inch diameter rigid copper tubing makes a good seal press tool. Clean the burrs and smooth the end of the copper tubing before using to prevent cutting the seal.

10. Remove the seal protector from the motor shaft key-way slot.

11. Install a new stainless steel snap ring, part number 718339 in the motor shaft groove. Use Tru-Arc pliers, see Figure 61.

12. Shims are provided in the seal package. The correct clearance from the top of the drain impeller blades to the drain cover bottom with the motor shaft pulled up should not exceed .020" or be less than .010", see Figure 65.

13. Proper clearance can be obtained by placing a 1/8 stack of washers over the shaft as shown in Figure 65 and temporarily installing the impeller over the shaft.

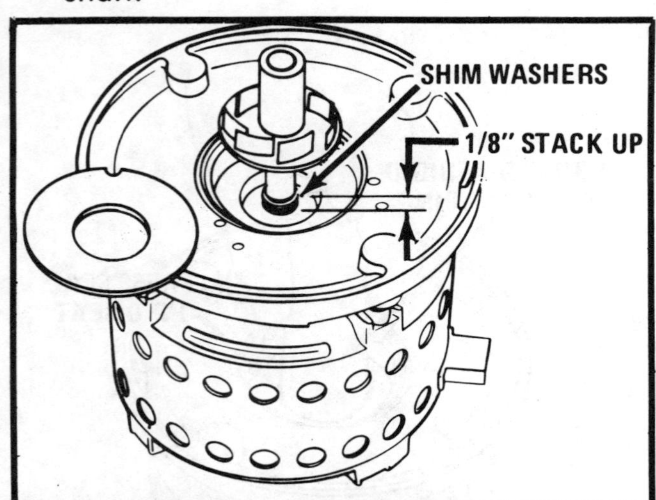

Figure 65

14. Install drain pump cover. Adjust shim washers stack up until the impeller just barely touches the cover while pulling up on the motor shaft.

15. When this point is established, remove one .015" shim; again install the impeller and drain pump cover.

16. Set pump housing so that large opening in the drain outlet is in line with the raised extension of the pump mounting plate. The housing is keyed to the pump mount by three locating pins.

17. Replace mounting screws and secure pump housing.

18. Replace locking ring part number 718342, position properly, Figure 66.

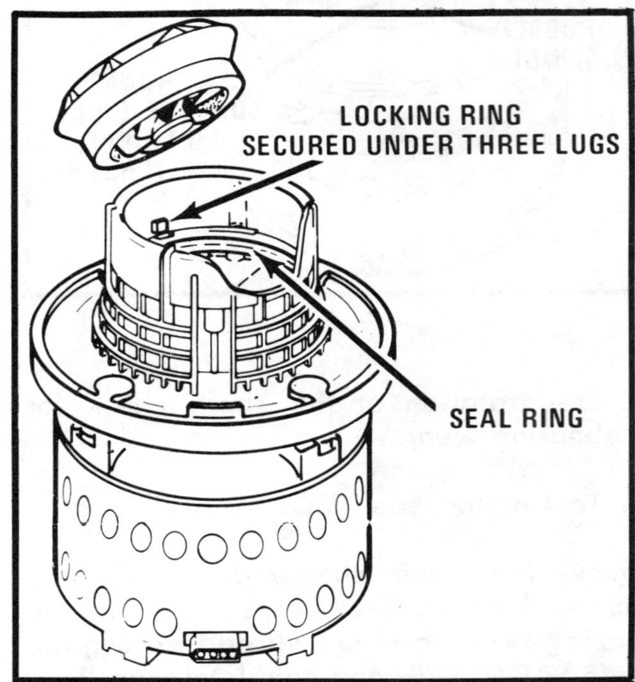

Figure 66

19. Install wash impeller assembly making sure that the stainless steel Axial impeller is mounted properly under the wash impeller, see Figure 67.

20. Secure the wash impeller on motor shaft with the impeller mounting screw and bend locking tab over the head of the screw after tightening the screw.

21. Manually spin impeller by hand and check for freedom of movement; there should not be any scraping sound, and shaft and impeller should turn smoothly.

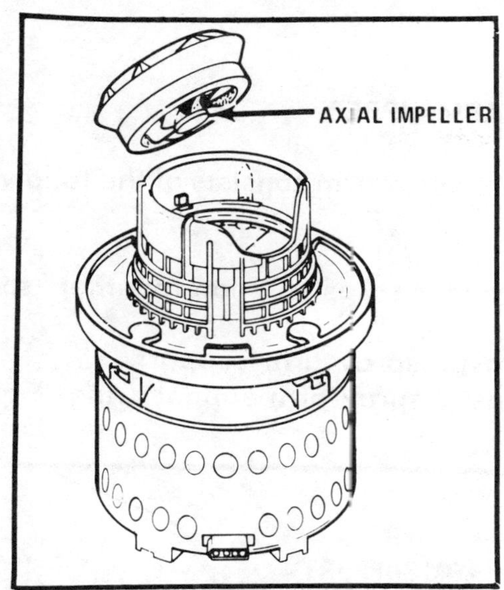

Figure 67

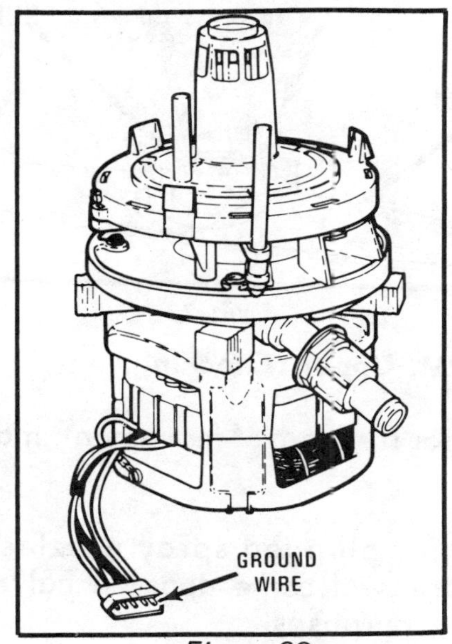

Figure 68

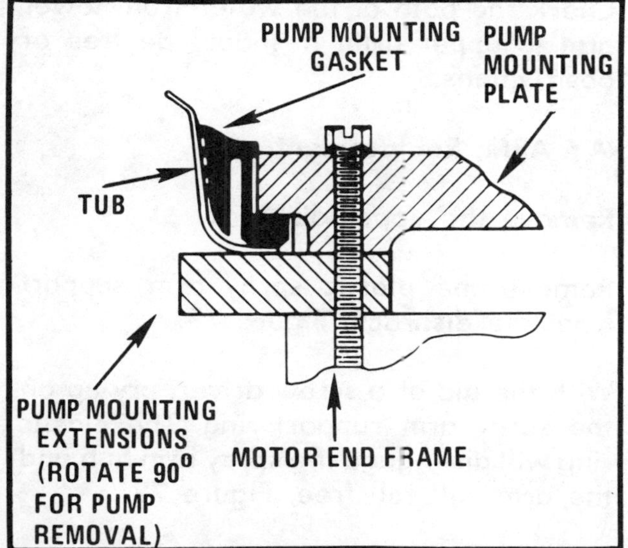

Figure 69

22. Position diffuser assembly into the locating slots in pump housing and secure by rotating the diffuser clockwise until locked. Insert the diffuser lock into place in any of the slots between the diffuser and the pump housing. Secure the locking tab behind the vertical extension of the housing, Figure 56.

NOTE: A new Whirlpool pump and motor assembly was introduced and used after June 1971. This pump assembly was and is used on all models after 1971. This is a centrifugal type pump and uses the open type motor (cube shaped), Figure 68. A new method of holding the pump in place is also used. Instead of the customary 4 metal clamps, 4 rubber hold downs are used. The hold-downs are rotated 90° to allow the pump and motor assembly to be removed, Figure 69.

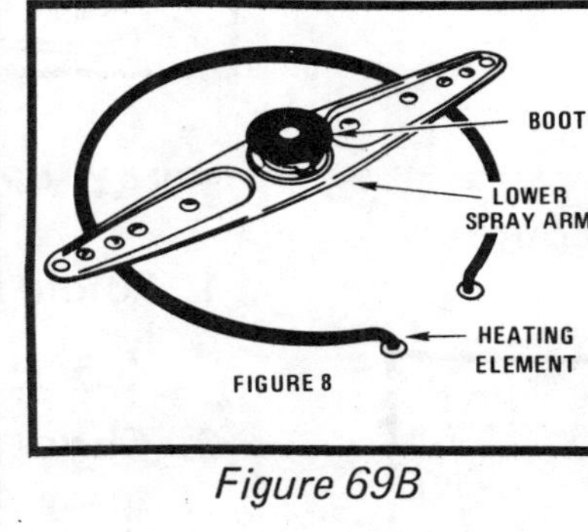

Figure 69B

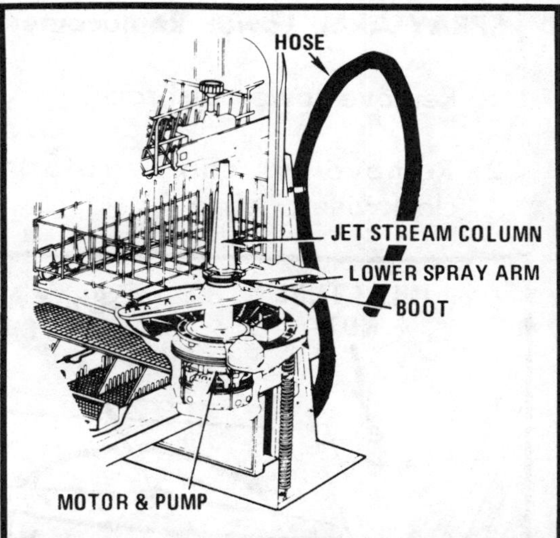

Figure 69A

JET STREAM COLUMN

A taller Jet stream column replaces the previously used tower. This prevents dishes from being loaded above the tower and blocking the water to the upper spray arm. Refer to *Figure 69A.*

NOZZLE AND BOOT ASSEMBLY

A rubber boot has been added to the top of the lower spray arm nozzle cap on all but the single level spray arm model SDU 3000, illustrated in *Figure 69B.* This boot seals against the bottom surface of the Jet Stream Column, thus delivering more water to the upper spray arm.

SPRAY ARM, Lower.

See text, "WATER CIRCULATION" for method of checking.

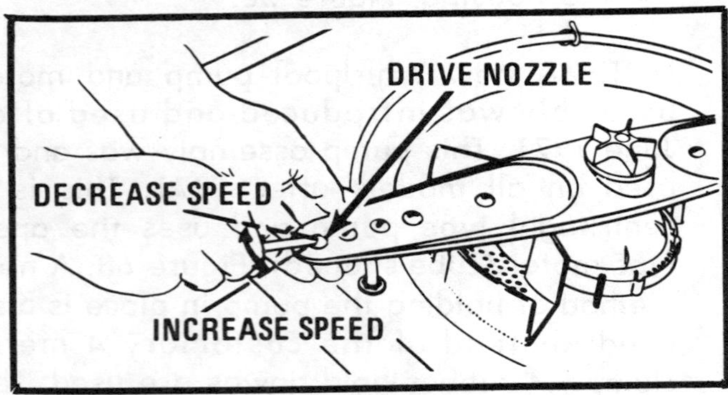

Figure 70

1. Spray arm can be adjusted by redirecting drive nozzles at each end of the arm, Figure 70.

2. Check for plugged spray nozzles.

SPRAY ARM, Lower Replacement.

1. Remove lower dishrack.

2. Remove hub cap by rotating counter-clockwise, Figure 71.

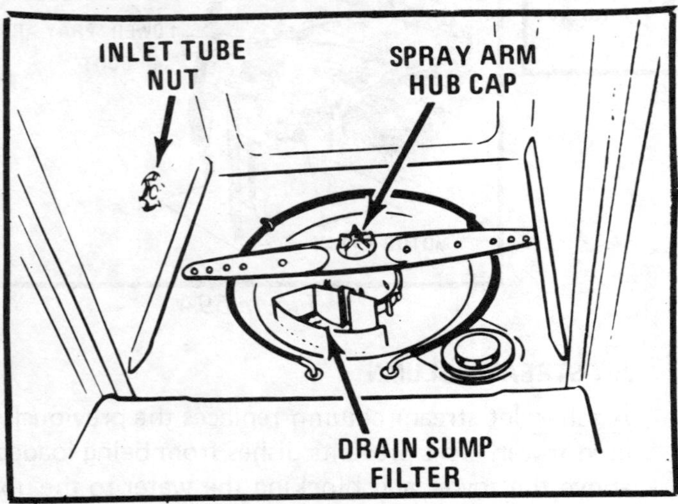

Figure 71

3. Remove spray arm.

4. To reinstall, reverse procedure.

SPRAY ARM UPPER

The upper spray arm consists of the following parts:

a. A two piece plastic "snap together" spray arm.
b. A plastic spray arm support.
c. A plastic spray arm support ring, Figure 72.

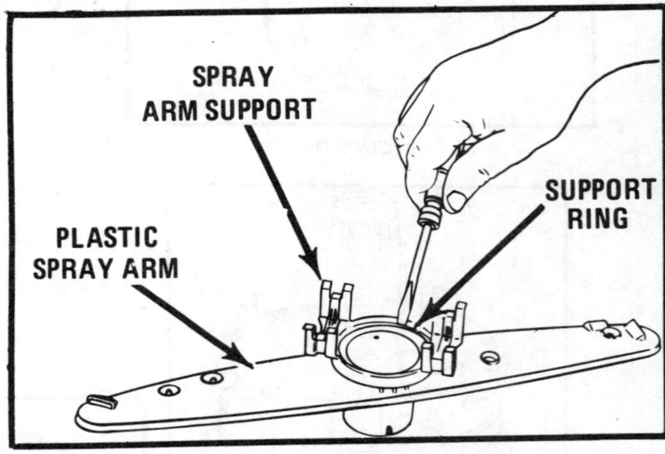

Figure 72

SPRAY ARM, Upper, Checking.

1. Rotate for freedom of movement, manually.

2. Check for plugged spray nozzles; the spray arm will come apart in halves for cleaning purposes.

3. Check the path of the water from lower arm to upper arm. It should be free of obstructions.

SPRAY ARM, Replacement.

1. Remove the upper dishrack.

2. Remove the plastic spray arm support from the dishrack, Figure 73.

3. With the aid of a screw driver, pry up on the spray arm support ring. The plastic ring will disengage the spray arm hub and the arm will fall free, Figure 72.

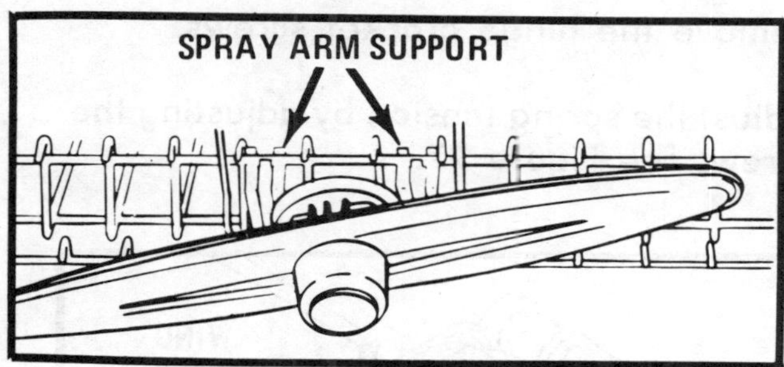

Figure 73

4. To reinstall, reverse procedure.

SPRAY ARM HUB

The spray arm hub is a molded plastic component and has a right hand thread that mounts the pump stator outlet port, Figure 71.

The spray arm hub should be visually checked for damage. A damaged hub will sometimes be the cause of the spray arm's failure to rotate. If replacement is necessary, rotate the hub counterclockwise for removal.

FILTER SCREEN

The filter is made of polypropylene and should be visually checked for damage. The filter should always be locked down into position by the three locking lugs, see Figure 74. Check for warping that may allow food through to the pump. The filter should always be checked for food particles clogging the screen.

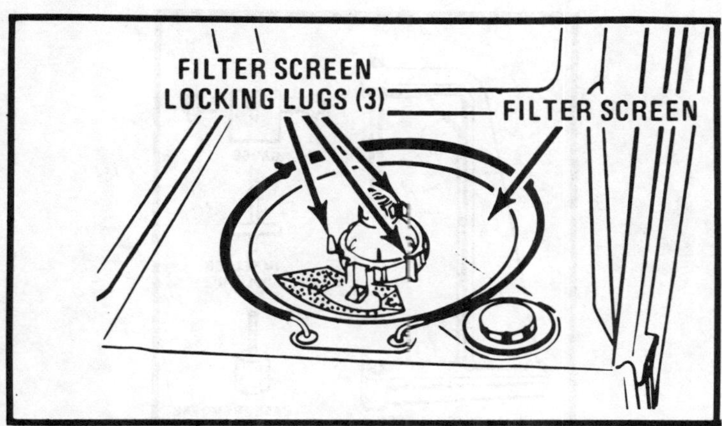

Figure 74

PORTABLE MODELS
Rear Shroud

1. Loosen or removen Allen set screw in timer knob and remove the knob.

2. Note Figure 75 and remove the Rear Shroud mounting screws.

3. Disengage the Rear Shroud from the bottom by lifting it up and over the timer shaft.

4. Remove the Rear Shroud from the dishwasher.

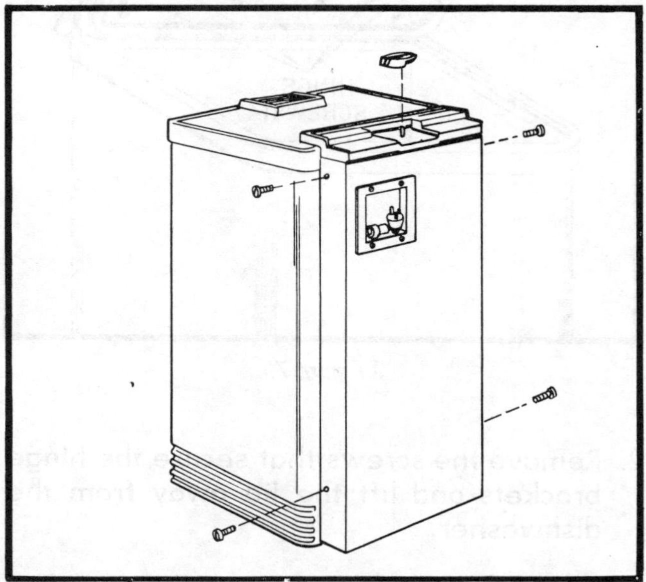

Figure 75

TUB GASKET

1. Remove the shroud.

2. Separate timer extension shaft, see "TIMER REMOVAL" text.

3. Lower the shaft extension; allow it to rest on the floor.

4. Raise the dishwasher lid and remove the two upper screws that secure each hinge assembly to the bracket to the rear of the dishwasher. Loosen the screws remaining but do not remove and pivot the lid assembly back from machine.

5. Remove the gasket; replace in reverse order.

LID, FP 20-40-50 Series.

The procedures in removing the lid will differ according to the model number.

1. With the lid open, remove the screws that hold the console lid, Figure 76.

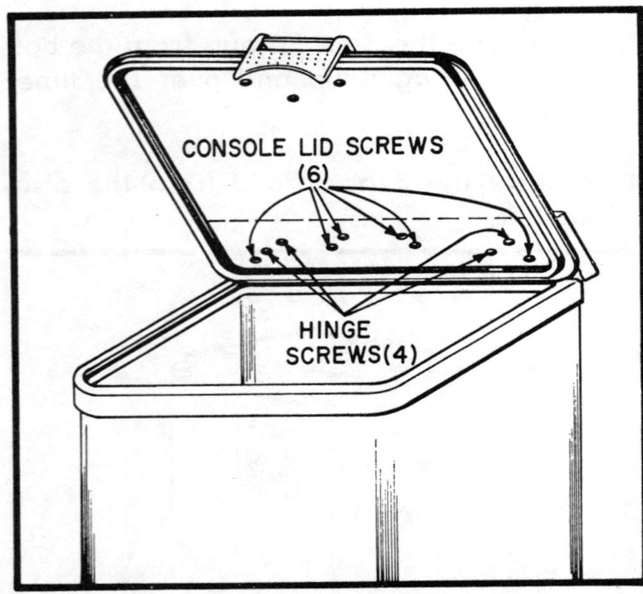

Figure 76

2. Remove the screws that secure the hinge brackets and lift the lid away from the dishwasher.

NOTE: If you are servicing a model FP50B series:

1. Lift the lid and remove the latch assembly screws (three).

2. Remove the screws, securing the trim at the edge of the lid, front and rear. Remove, trim and lift out the work panel.

3. Remove the two screws securing each hinge bracket. Lift the interior section of the lid from the dishwasher.

HINGE

1. Remove the shroud from the rear.

2. Remove the lid.

3. Remove the hinge bracket screws.

4. Adjust the spring tension by adjusting the screw, see Figure 77.

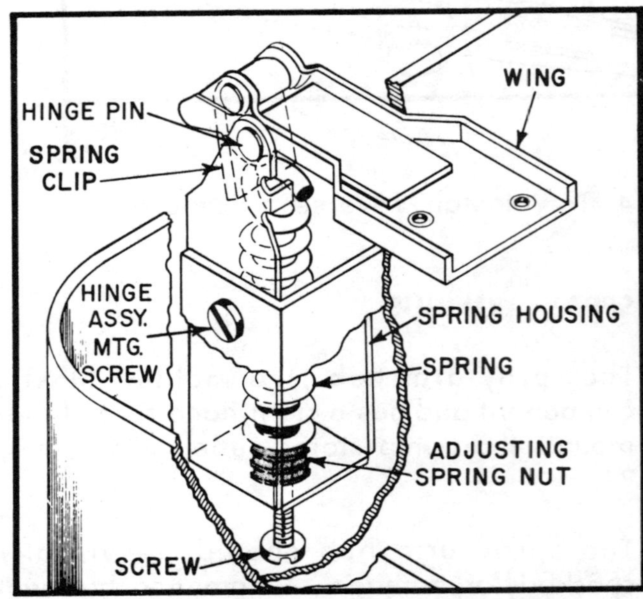

Figure 77

SEAL REPLACEMENT, Impeller Type Motor.

See text, "MOTOR REMOVAL, VERTICAL MOTOR."

When repairing a pump it is wise to always replace the seal assembly, to avoid possible damage to the pump motor resulting from a leaky seal.

The pump impeller service kit (Robinair No. 12863) is shown in Figure 78. Additionally, you will need an electric drill, a 9/32" drill bit, centerpunch and vise-grip pliers.

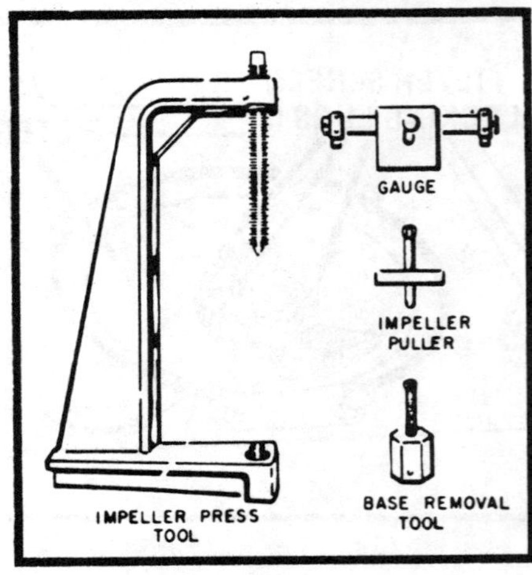

Figure 78

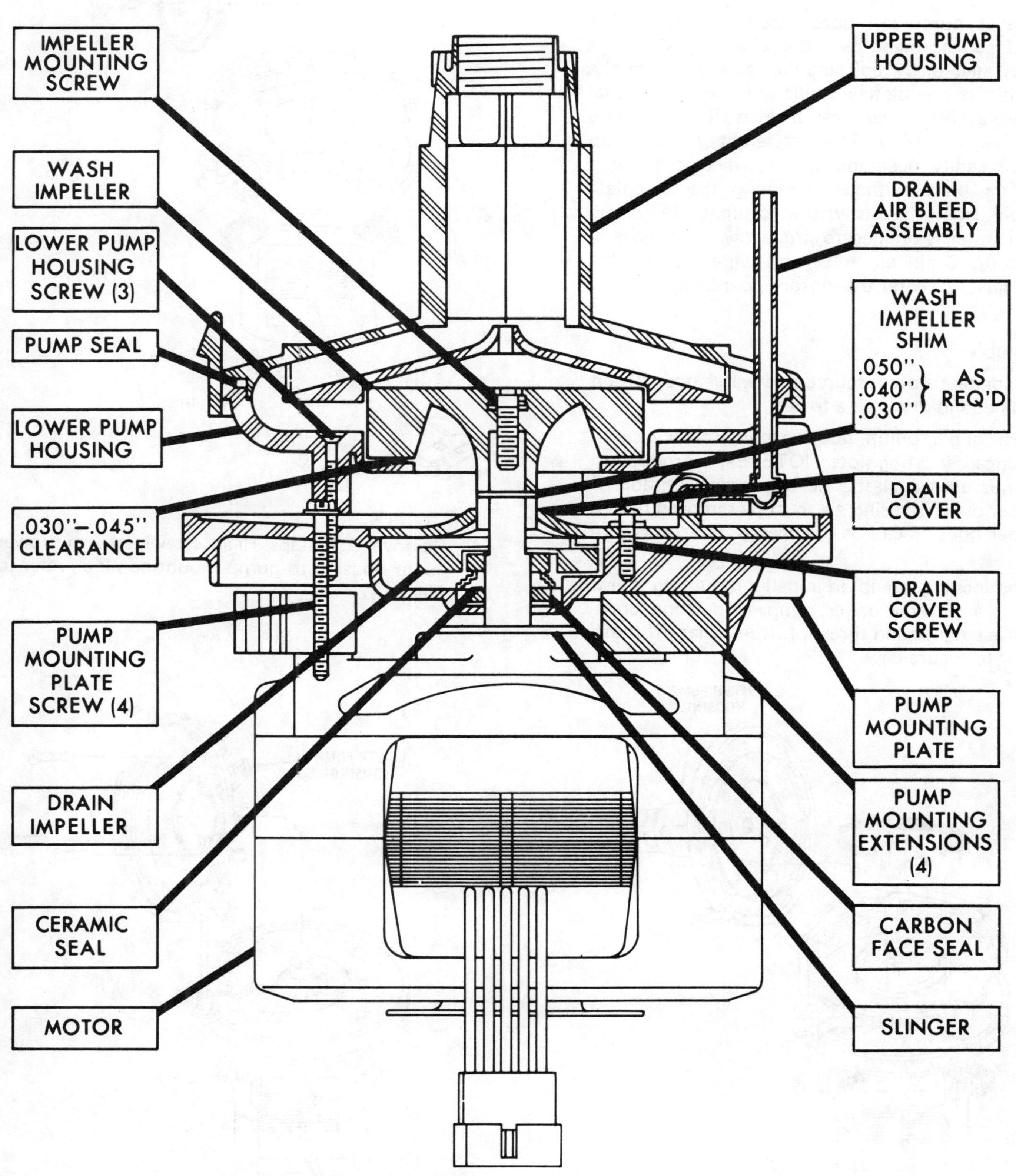

Figure 79, Whirlpool Pump and Motor Assembly — Cutaway View

PUMP AND MOTOR, PART NO. 718551

Repair Procedure

Repairs on pump and motor assembly part number 718551 can be made by either changing the part/s that has failed or by replacing the complete assembly. The unit is completely serviceable and individual parts are available. Parts can be identified in the cutaway view, Figure 79. The disassembly of the pump can be readily accomplished down to the pump mounting plate without removing the complete assembly from the dishwasher cabinet. Be certain that all water is drained to prevent water damage to the motor. Drain all hoses and wipe cabinet dry. Water must not enter the motor upper bearing.

Disassembly

1. Disconnect power source and water hose. Drain cabinet and dry with a towel.

2. Turn upper pump housing conterclockwise to disengage locating slots. NOTE: Early production pumps used a plastic tie to prevent rotation of upper pump housing; tie must be removed before upper housing can be disengaged, refer to Figure 80.

3. Bend locking tab up on impeller mounting screw. Using a 5/16 nut driver, remove mounting screw, this is a right hand thread. Lift impeller off shaft, refer to Figure 81.

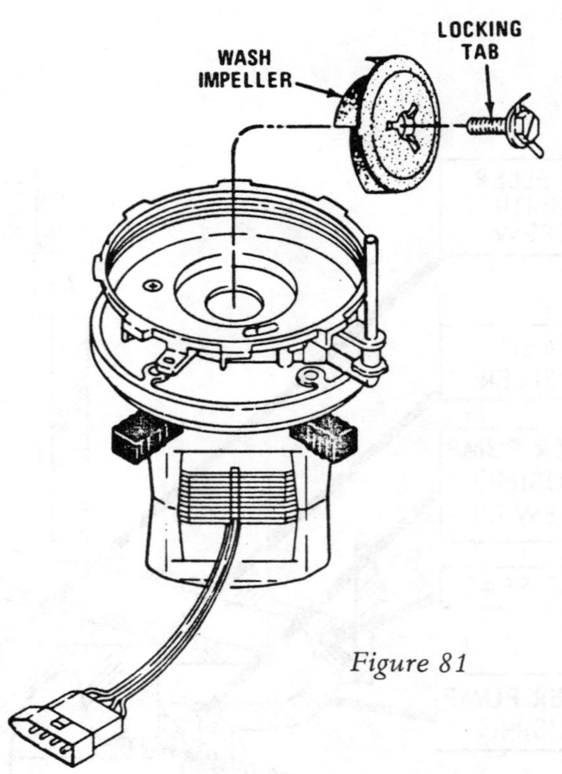

Figure 81

4. Remove 3 Phillips Head screws securing lower pump housing to pump mounting plate, refer to Figure 82.

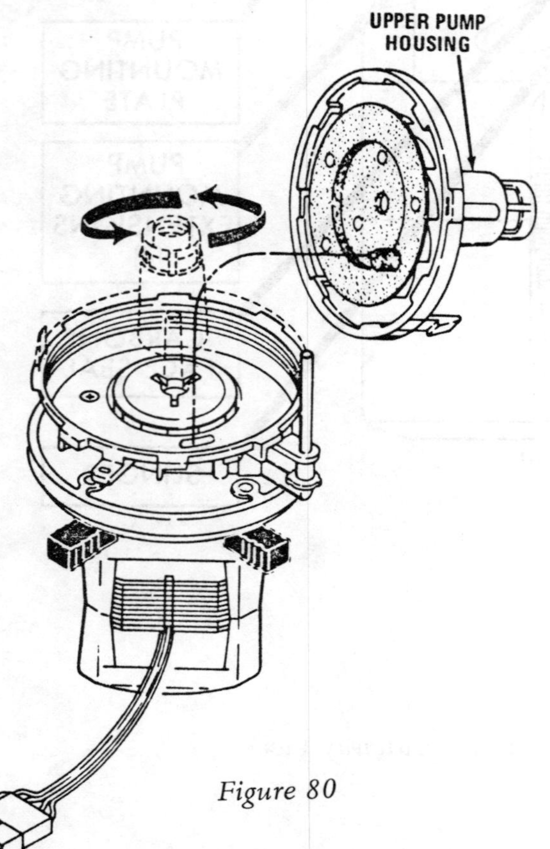

Figure 80

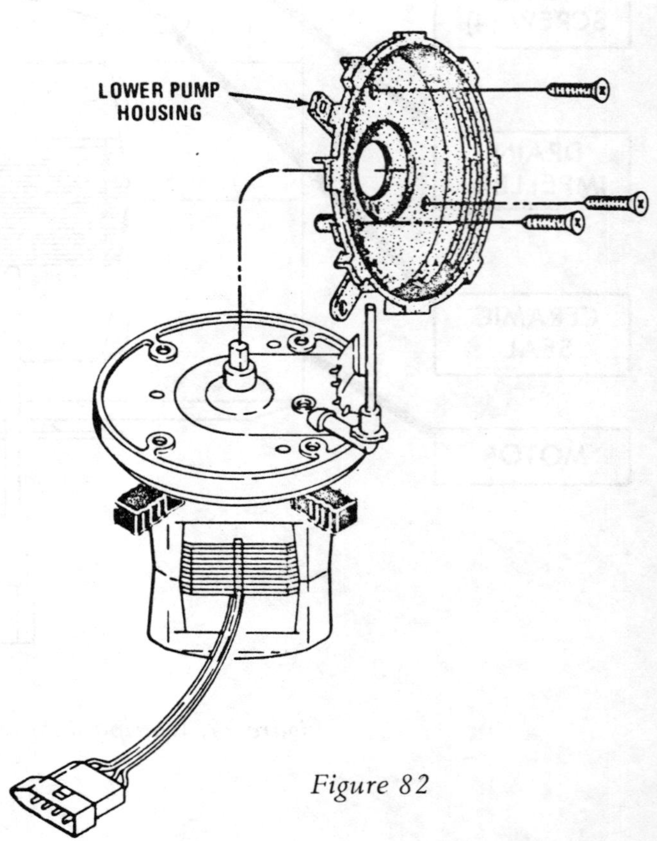

Figure 82

5. Remove drain pump cover screw, Figure 83, and remove drain pump cover and air bleed assembly.

6. Remove wash impeller shim/s, refer to Figure 84.

7. Pull straight up on impeller and remove. If motor is replaced shims will be on new replacement motor. Use shims as required, refer to Figure 85.

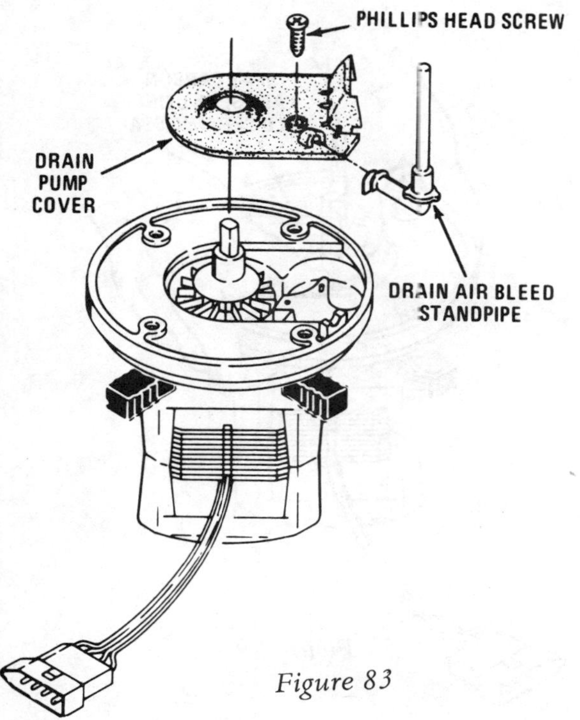

Figure 83

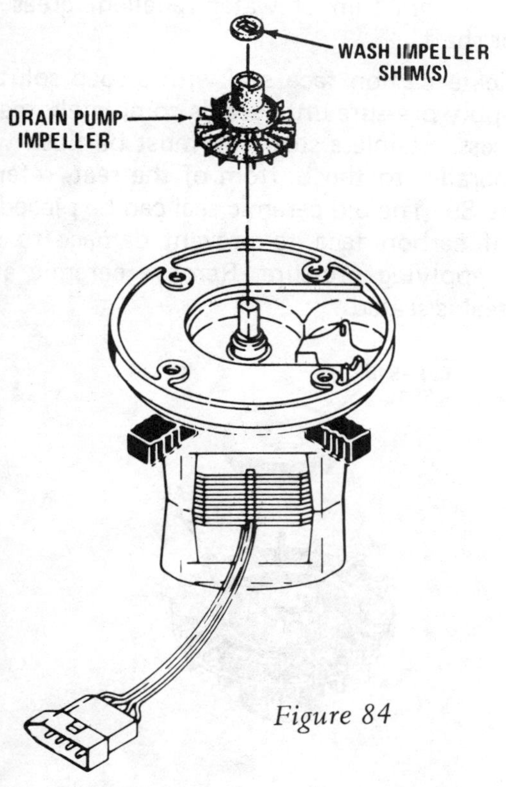

Figure 84

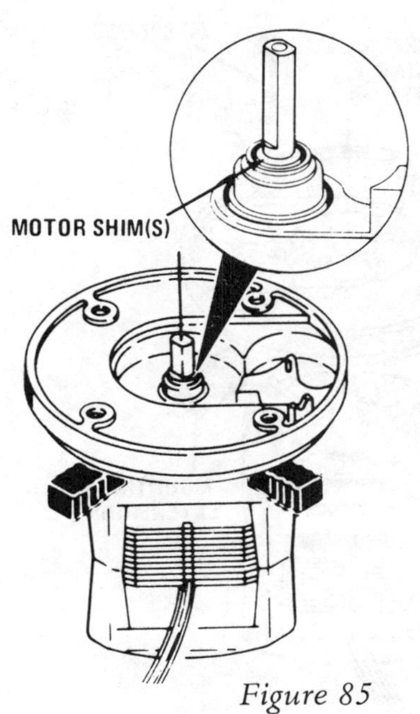

Figure 85

8. Both parts of the seal assembly are now visible. Both parts can be easily pried out to replace. When rebuilding pump, always install a new seal assembly and impeller. These parts are available in Kit No. 675153, refer to Figure 86.

9. Remove 4 screws securing pump mounting plate to motor. Lift pump plate assembly off motor, refer to Figure 87.

10. Rubber pump mounting extensions can now be removed from pump mounting plate, refer to Figure 87.

11. If motor replacement is necessary, slinger can now be removed, refer to Figure 87.

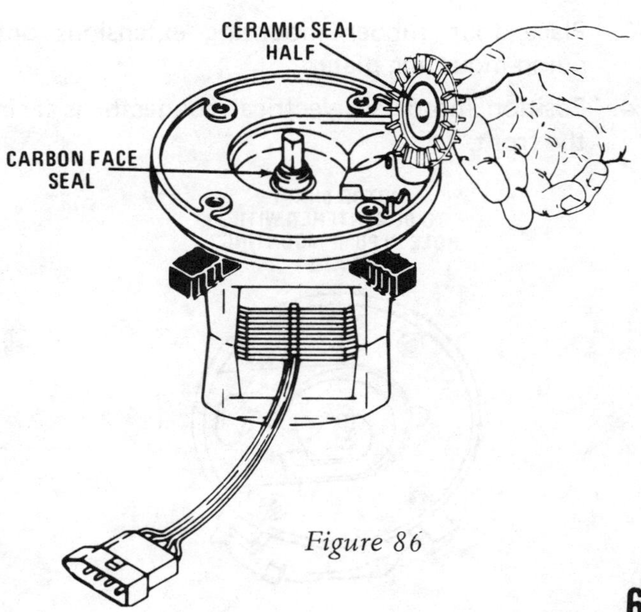

Figure 86

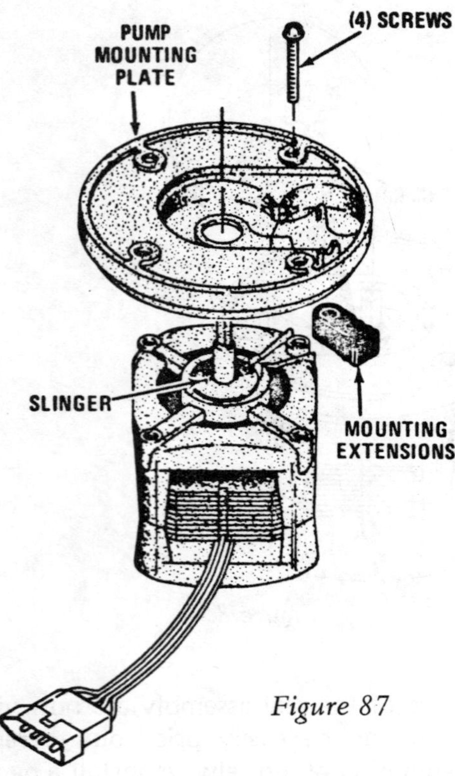

Figure 87

Reassembly: Before assembly is attempted, the motor must be checked for the following, Shaft end play —maximum is .010, motor should rotate freely, look for evidence of water leaking into the top motor bearing. Add a few drops of SAE #20 oil to the bearing resevoir. Check motor for correct operation, check with volt/ohm meter for grounded motor.

1. Install slinger, press onto motor shaft. It must be below the top edge of recessed water trough, but must not touch the bearing shield.

2. With a suitable cleaning fluid, wipe both surfaces of the seal assembly recesses, refer to Figure 86.

3. Place four rubber mounting extensions onto pump mounting plate.

4. Position motor so electrical connector is facing the front.

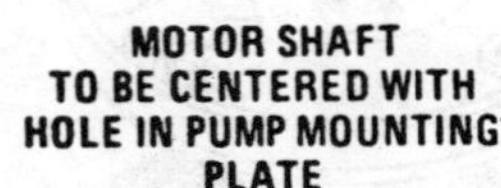

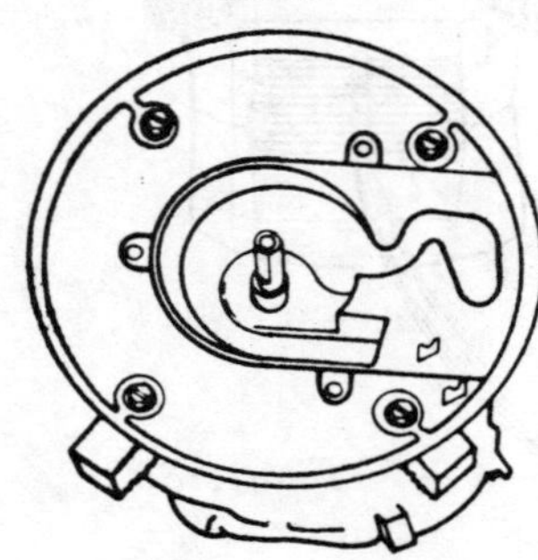

Figure 88

5. Position pump mounting plate on motor so drain outlet is facing right side.

6. Install motor mounting screws through plate, do not completely tighten the 4 mounting screws.

7. Center pump mounting plate in relation to motor shaft. Finish tightening the 4 mounting plate screws securing the motor, refer to Figure 88.

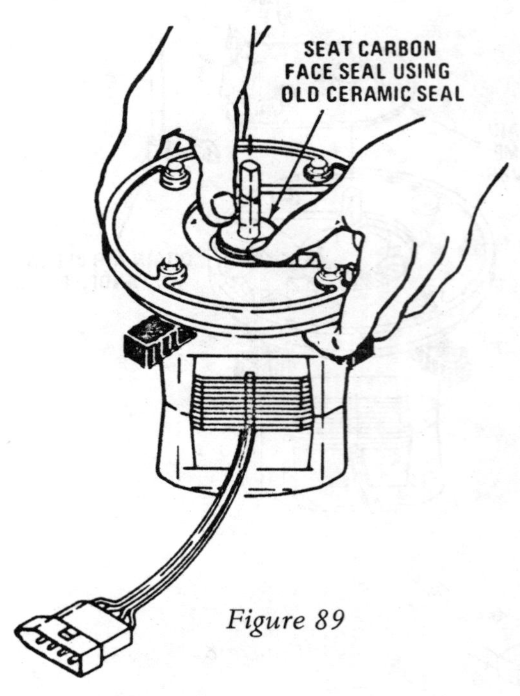

Figure 89

8. Apply a thin film of water repellent grease to motor shaft.

9. Lubricate carbon face seal with a soap solution and apply pressure until seal is completely seated in recess. Stainless steel cup must be flush with and parallel to the bottom of the seat, refer to Figure 89. The old ceramic seal can be placed on top of carbon face to prevent damage to seal when applying pressure. Remove ceramic after new seal is seated.

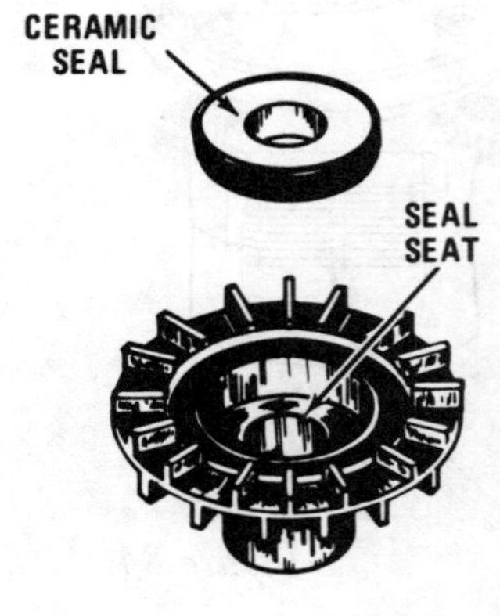

Figure 90

10. Lubricate new ceramic seal assembly with a soap solution, press firmly into the recess provided in the drain impeller, refer to Figure 90.

11. Install drain impeller over motor shaft to seal surfaces mate. Proper spacing has been assured by the height of the motor shaft shoulder so it is not necessary to check drain impeller clearance.

12. Install drain air bleed standpipe assembly to drain cover. Secure with Phillips Head screw, refer to Figure 91.

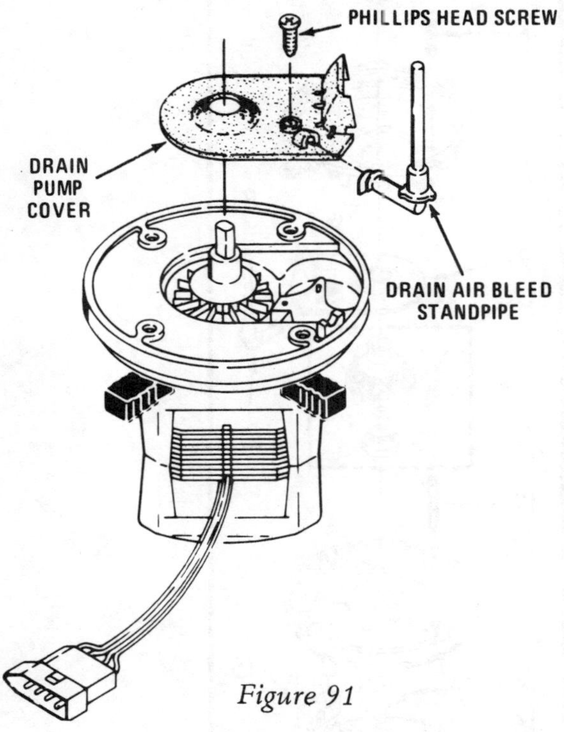

Figure 91

13. Install lower pump housing to pump mounting plate with 3 Phillips screws, refer to Figure 82 in Disassembly Procedure.

14. Place wash impeller on motor shaft. Apply pressure downward to assure seat is against shaft shoulder. Check clearance as illustrated in Figure 92. A wire gauge is included in kit no. 675153. Adjust shim stack to allow proper clearance between .030 inch to .045 inch. Before securing drain impeller, apply water repellent grease to both sides of all impeller shims.

15. Tighten wash impeller screw and bend down locking tab. Reassemble upper pump housing to lower pump housing.

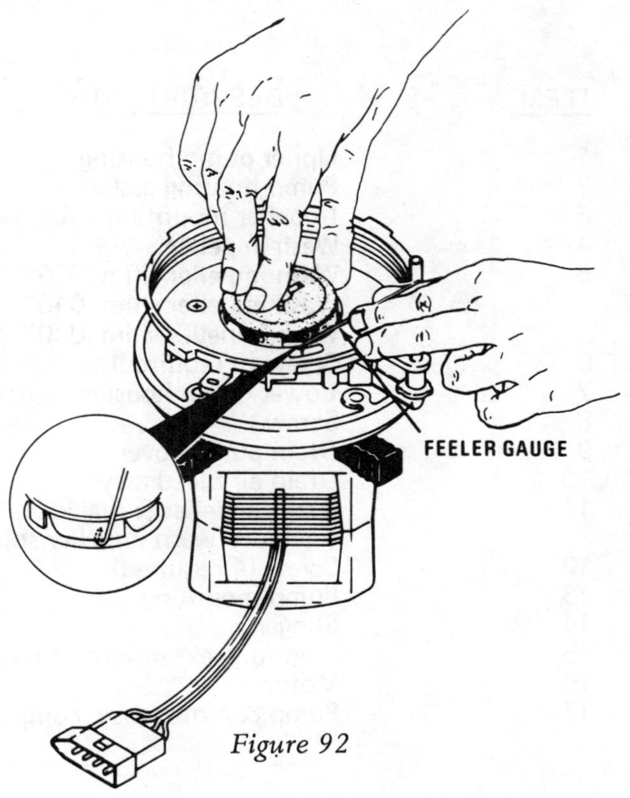

Figure 92

REPLACEMENT PARTS

Figure 93 is a parts breakdown for Part No. 718551. When ordering parts be sure to specify the correct model and serial number of the dishwasher.

All dishwasher pump and motor assemblies are ink stamped to insure proper replacement. The part number appears on the motor laminations. Refer to Figure 94. Part number for the motor only is still found on the end bell of the motor.

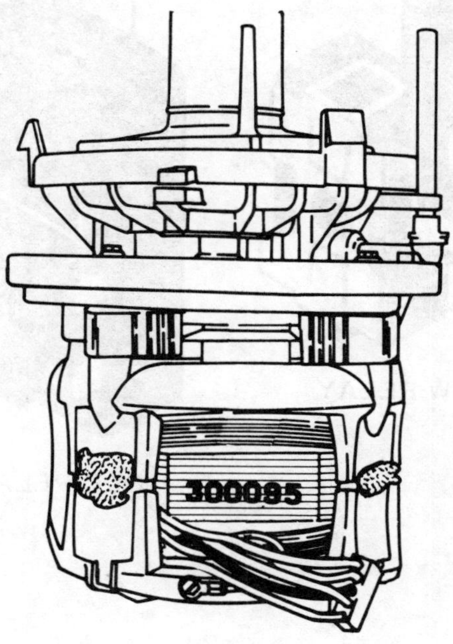

Figure 94

ITEM	DESCRIPTION
1	Upper pump housing
2	Pump housing gasket
3	Impeller mounting screw
4	Wash impeller
5	Wash impeller shim .050″
	Wash impeller shim .040″
	Wash impeller shim .030″
6	Screw (3 required)
7	Lower pump housing
8	Screw
9	Drain pump cover
10	Drain air bleed assy.
11	Drain impeller & seal kit (includes wash impeller shims)
12	Screw (4 required)
13	Pump mounting plate
14	Slinger
15	Mounting extensions (4 required)
16	Motor
17	Pump & motor assy. complete

GRAVITY RELAY KIT

Whenever a pump motor assembly is replaced on a dishwasher having a console mounted realy, the relay must be replaced. The replacement relay Kit Number 675262 will replace 719364, 675216 and 719958. The kit includes adapters for mounting and an instruction sheet for installation.

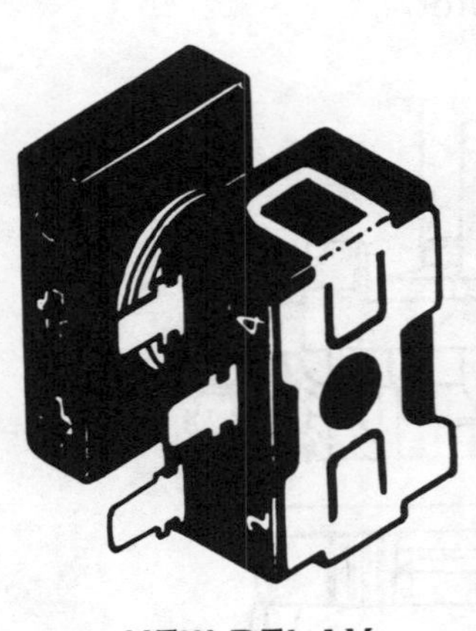

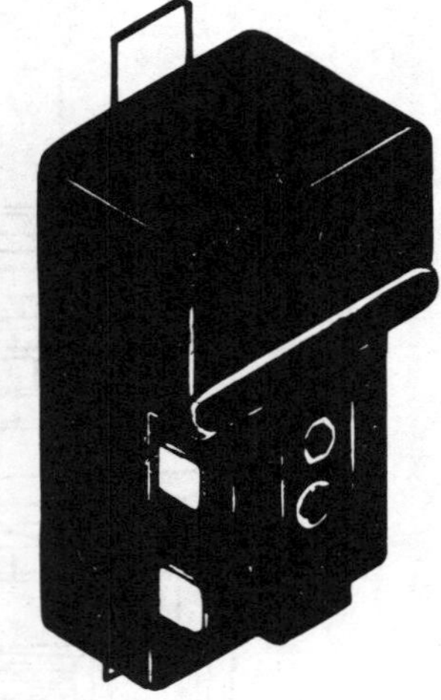

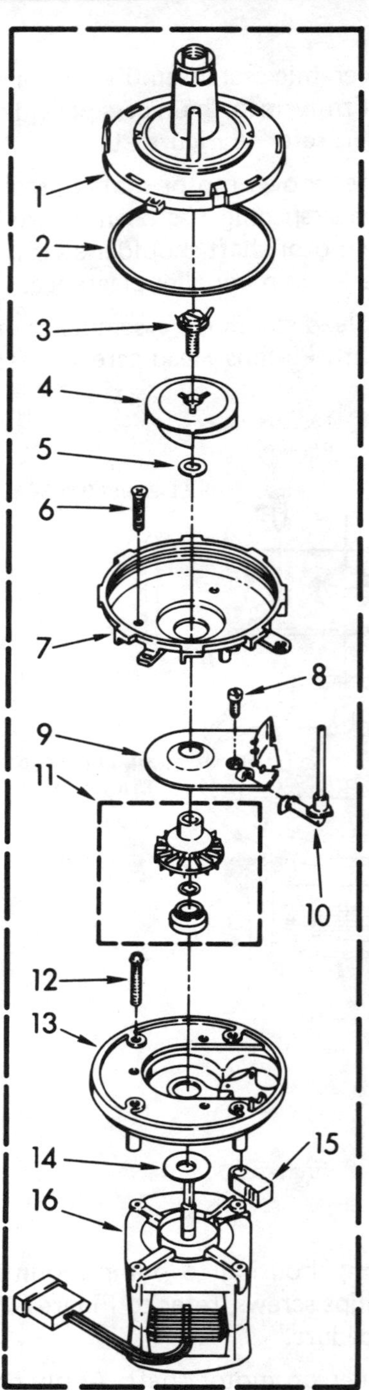

Figure 93, Parts Breakdown

PLUMBING REQUIREMENTS

NOTE: Observe all local plumbing codes.

a. Water supply — 15 pounds minimum and 120 pounds maximum water pressure. Pressures in excess of 120 pounds will require a pressure reducing valve installed in the primary supply line.

b. Water temperature must be within 140°
and 160° F. Cooler water will affect the
washing quality of the dishwasher.

c. Figure 95 illustrates a faucet coupler
which is attached to the hoses on portable
model dishwashers. There have been several types of couplers used since production started. All model FP-50 and model
FP-40 dishwashers use a duplex coupler
where both intake and outlet hoses are
contained in the coupler assembly. Model
FP-20 series have a single connection to
the faucet with a remote drain hose. All
the couplers require an adapter on the
faucet to accommodate the snap-on feature. Although the adapters do vary, the
procedure for installing them is similar.
This procedure is contained in the following paragraph and in the installation
instruction sheet.

d. If the faucet has either internal or exter-
nal threads, remove the aerator from the
faucet and thread the appropriate faucet
nipple and gasket on the faucet. These
parts may remain on the faucet when the
dishwasher is not connected.

e. If the faucet spout **does not** have threads,
special parts are required to connect the
dishwasher supply hose to the faucet,

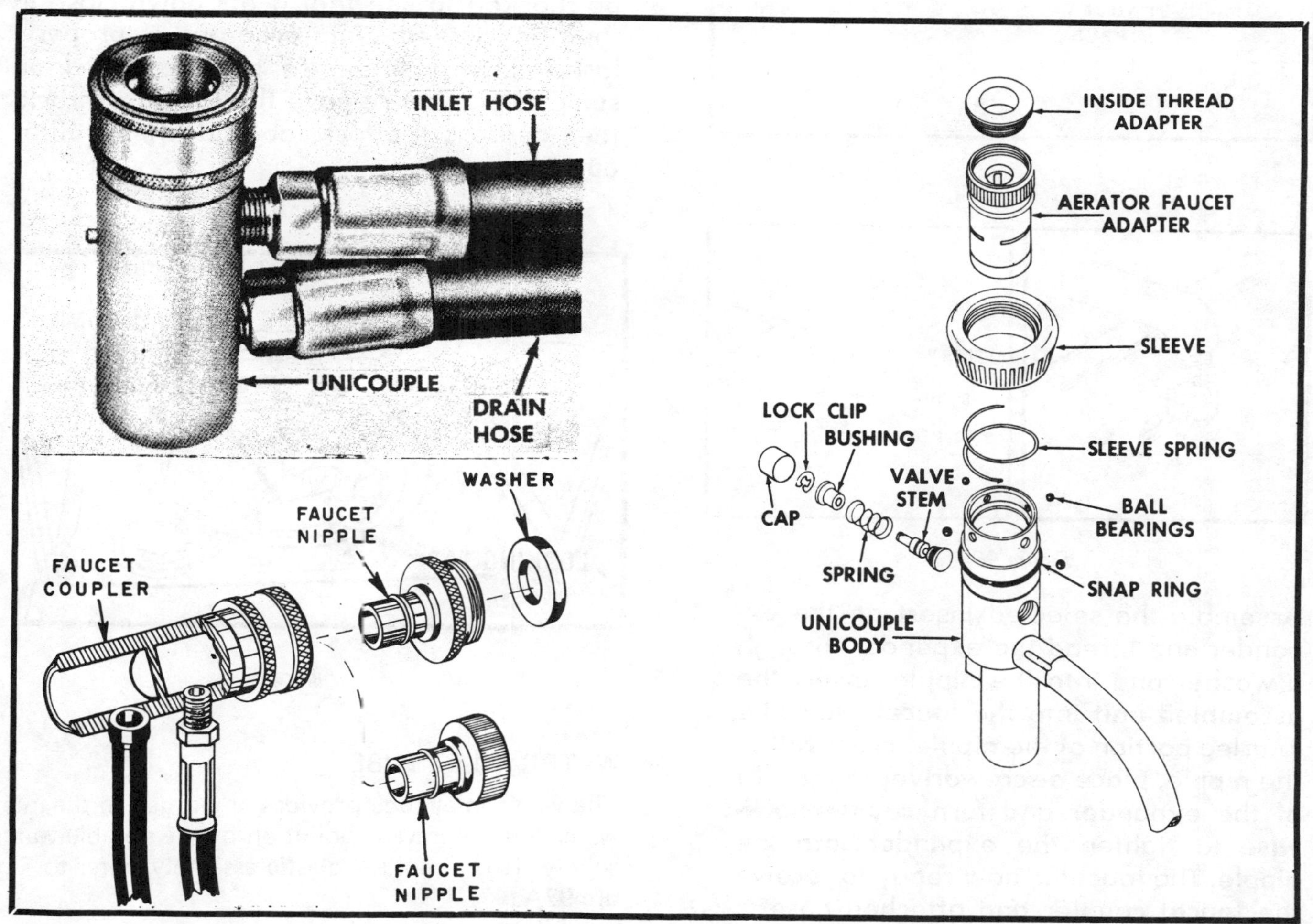

Figure 95

Figure 96. Select an insert that is approximately the same diameter as the inside diameter of the faucet. The hole in the insert is chamfered so that the opening on one face or side of the insert is larger than the one on the other. The tapered head on the expander fits snugly into the chamfered hole. When tightened, it causes the insert to expand.

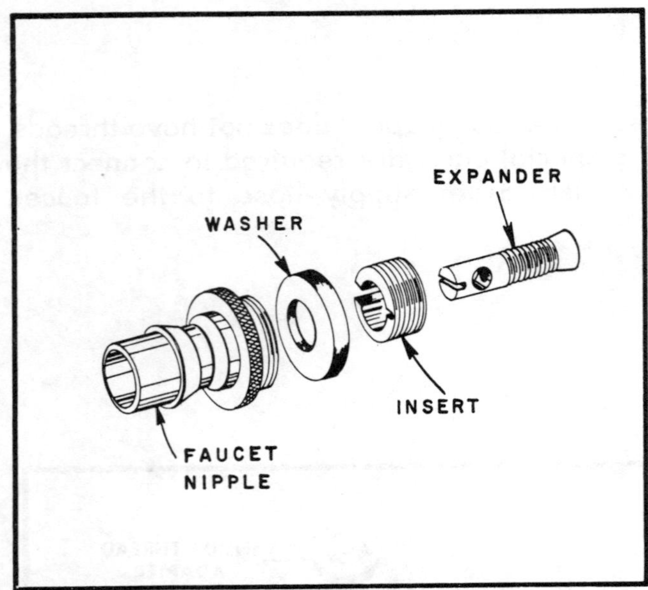

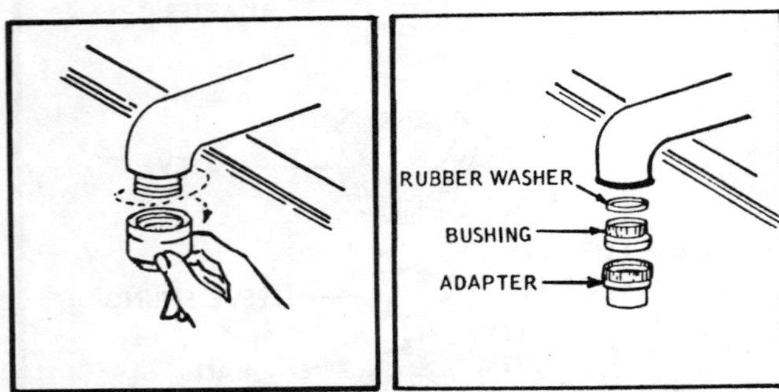

Figure 96

Assemble the selected insert on the expander and thread the expander through a washer and into the nipple. Insert the assembled unit into the faucet. Hold the knurled portion of the nipple. From within the nipple, place a screwdriver in the slot of the expander and turn **counterclockwise** to tighten the expander into the nipple. The faucet is now ready to receive the faucet coupler and attached hose.

f. To attach the faucet coupler, retract the spring-loaded knurled ring on the faucet coupler. Position the coupler on the protruding faucet nipple and release the ring. Enough force is exerted to keep the coupler secure. To remove — turn water supply off, retract the ring and disengage the coupler.

g. The drain hose should be positioned so that it discharges into the sink or other drain equipped bowl. Any standard sink or tub drain is sufficient to handle the discharge flow.

DRAIN SUMP FILTER

The drain sump filter is made of molded polypropylene and locks into the filter screen by the use of plastic tabs. Its purpose is to prevent large pieces of food from entering and damaging the drain pump and possibly plugging the drain hose and system. It should be checked to see that it fits down close in the filter screen sump area and is properly locked into position. To remove the drain sump filter, first remove the lower dishrack, then depress the lock tab and lift the filter out, Figure 97.

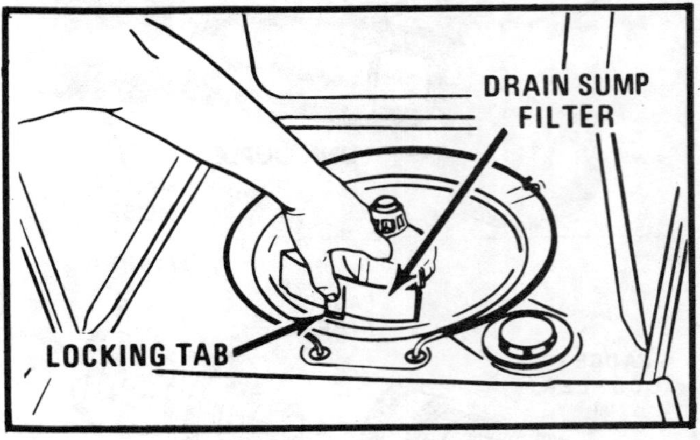

Figure 97

WATER INLET TUBE

The water inlet tube provides an air gap in the inlet water line to prevent pollution of the potable water supply. It is a molded plastic assembly, refer to Figure 97A.

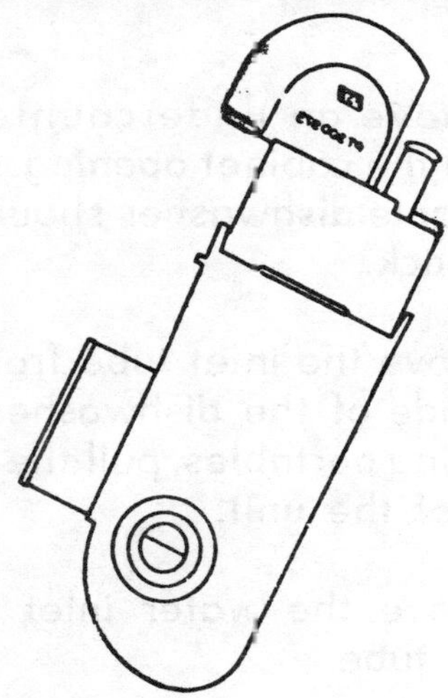

Figure 97A, Water Inlet Tube

THE WATER VALVE, Early and Late Models.

The water inlet valve is solenoid operated. When an electric current passes through a coil of wire, it produces a magnetic field. This principle is used to operate the water inlet valve.

FLOW WASHERS

If the proper quantity of water is not being metered into the dishwasher, it is possible the flow washer has become worn or deteriorated with age. To replace the flow washer, remove the water inlet tube, the brass fitting, Figure 108, and remove the flow washer. When a new flow washer is installed be sure the identifying numbers or letters on the flow washer are towards the incoming water, see Figure 109A.

WATER INLET

The water connection to the Portable as well as the Undercounter Models comes off of the hot water supply. Prior to operating the dishwasher it is advisable to turn the hot water tap on in the kitchen sink until the hot water comes out hot. If the hot water heater is some distance from the dishwasher this is especially meaningful, as the initial fill of water in the dishwasher could possibly be just luke warm. While the portable dishwasher is equipped with a "Quick Couple" adapter to connect to the kitchen faucet, the Undercounter models must be connected to the plumbing with copper tubing. The hot water supply should be between 140° to 160° to assure maximum results in dishwashing.

WATER DRAIN

On Portable Dishwashers, a drain hose is provided as a means of discharging water from the pump to any standard sink drain. Undercounter Dishwashers require a permanent (code approved) drain connection to a drain receptacle or drain line. It is recommended that an Air-Gap be installed in the dishwasher drain line for undercounter installation. In Figure 110 is shown the Gemline AF-4 deck mounted Air-Gap available at your appliance parts supplier.

The following should be checked:

1. Proper assembly to the dishwasher cabinet, see Figure 98.

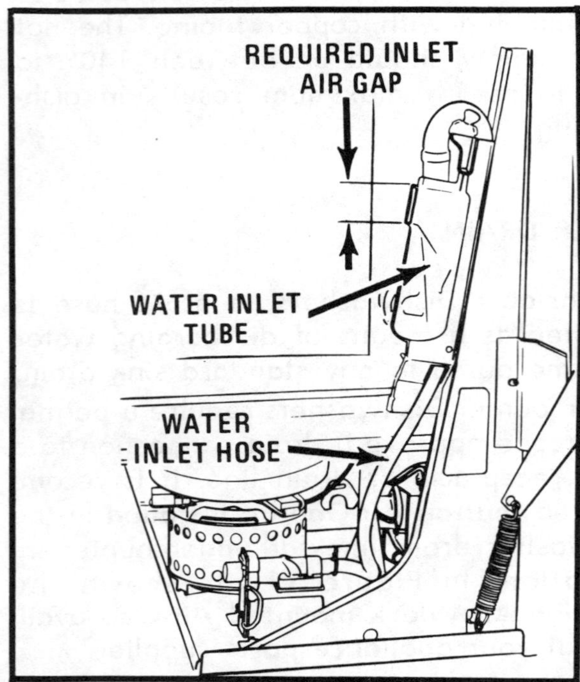

Figure 98

2. Check for leaks along the welded seam.

3. Check water outlet port for mineral deposits. An accumulation of these deposits will cause the water to spray incorrectly.

4. Check for excessive splash through the air gap.

WATER INLET TUBE, Replacement.

1. Remove the lower dishrack.

2. Remove the plastic nut that holds the inlet tube to the inside of the dishwasher, Figure 71.

3. Remove an undercounter dishwasher from the cabinet opening. A front loading portable dishwasher should be placed on its back.

4. Remove the inlet tube from its mounting on side of the dishwasher tub. On front loading portables, pull the tube down and out of the unit.

5. Remove the water inlet hose from the inlet tube.

6. To reassemble, reverse procedure; take note of gasket positioning, Figure 99.

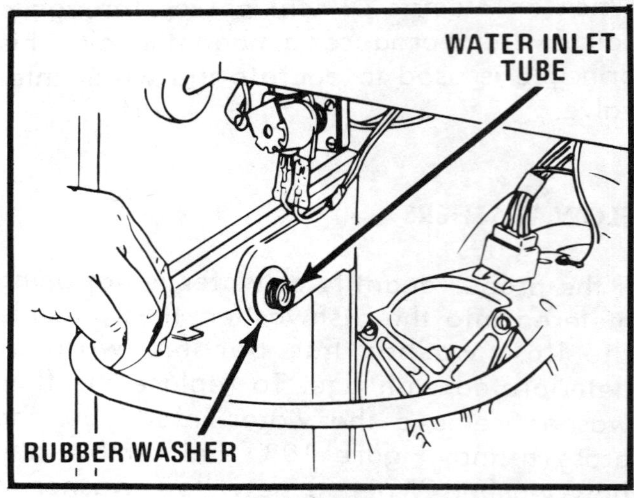

Figure 99

DRAIN LOOP

The drain loop is used on undercounter installations only. Its purpose is to prevent water syphoning from the dishwasher during the wash or rinse periods. It is positioned on the right side of the tub by a wire hose retainer and the two tie straps, Figure 100.

DRAIN LOOP, Checking

1. Check for restrictions such as kinks, food particles and waste material. If the drain is connected to the disposal, remove the connection at this point and check both

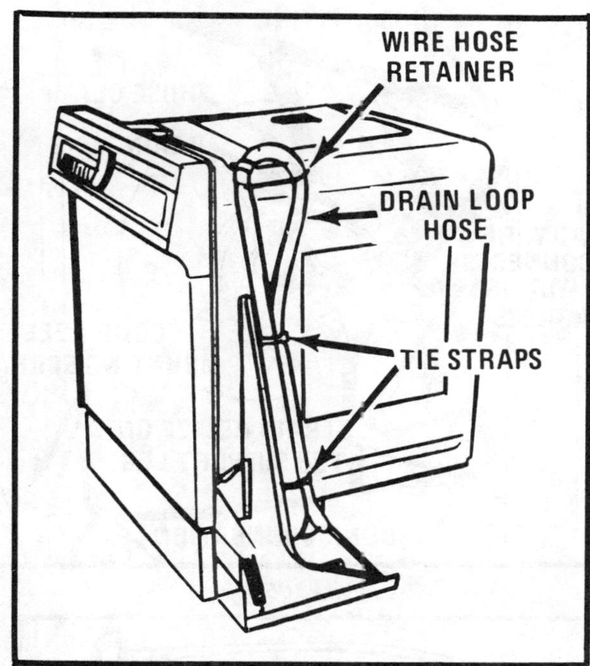

Figure 100

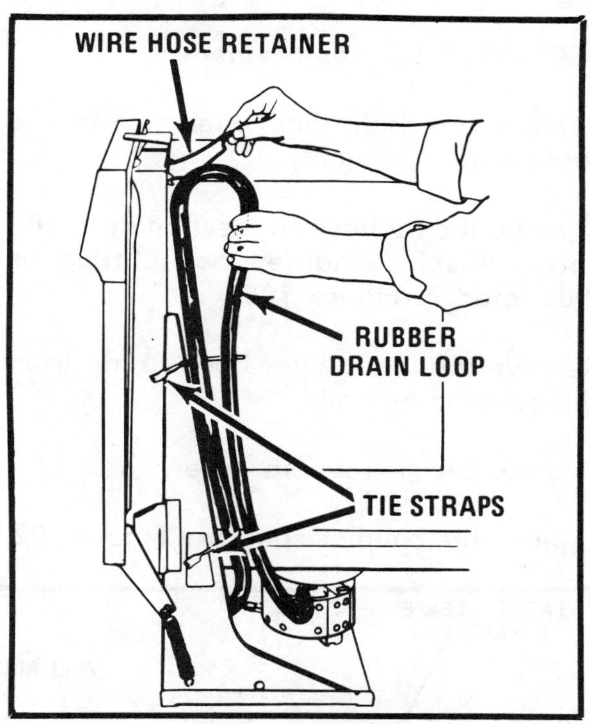

Figure 101

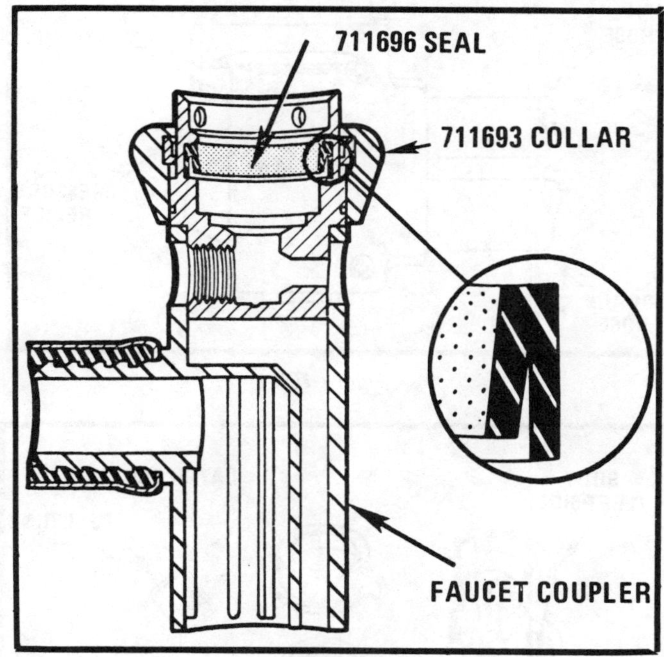

Figure 102

ends; check both the disposal outlet and the hose. This is the most likely place for a restriction.

DRAIN LOOP, Replacement.

1. Remove the dishwasher from the cabinet opening.

2. Disconnect the drain loop at the drain pump connection. Be prepared to catch the remaining water in the hose.

3. Lift up the wire hose retainer, Figure 101.

4. Remove the straps and lift the drain hose loop free of the dishwasher.

FAUCET COUPLER

Replacement of the faucet coupler necessitates replacement of the complete hose assembly. The locking collar and coupler seal are available as separate replacement items, Figure 102.

NOTE: All SXF models have a new type of faucet coupler which incorporates a relief button similar to the previously used coupler, Figure 103. Figure 104 illustrates the parts breakdown of the new faucet coupler.

FAUCET COUPLER, Replacement.

1. Remove the drain line connection at the drain pump assembly.

2. Remove the fill hose connection at the fill valve; this is a garden hose fitting as illustrated in Figure 105.

3. Remove the inlet water hose fitting from the coupler assembly.

4. Reassemble in reverse order.

To replace the coupler seal, see Figure 102.

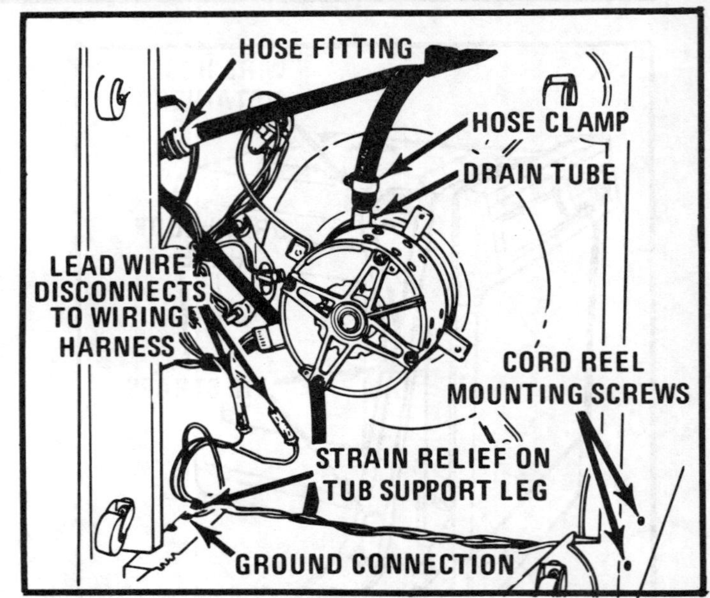

Figure 105

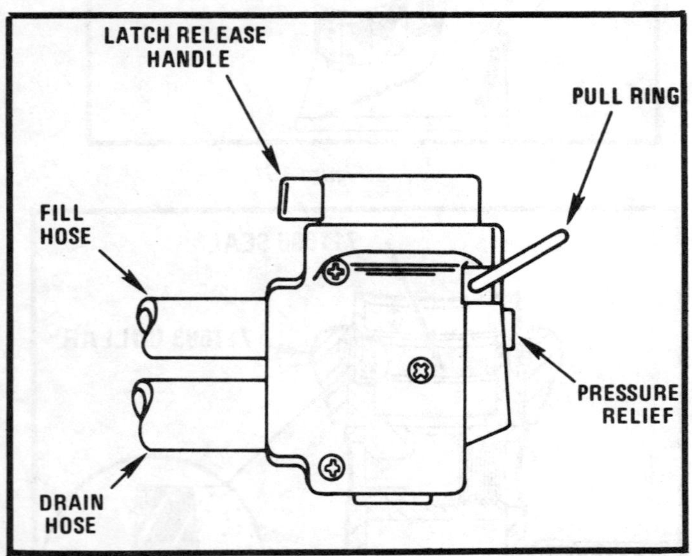

Figure 103

Figure 106

When the fill hose and/or the drain hose and faucet coupler are to be replaced the following procedure should be followed. Remove the fill hose from the fill valve and the drain hose from the pump adapter. Unstrap the hoses from the lower hose bracket. Remove the dishwasher top, see text. Pull the hoses out of the hose housing and remove the remaining two straps on the hoses. Reassemble in reverse order.

FILL AND DRAIN HOSE ADJUSTMENT

1. Place the hose assembly on a flat surface in the configuration as shown in Figure 106.

2. Tie the straps in place as shown in Figure 106.

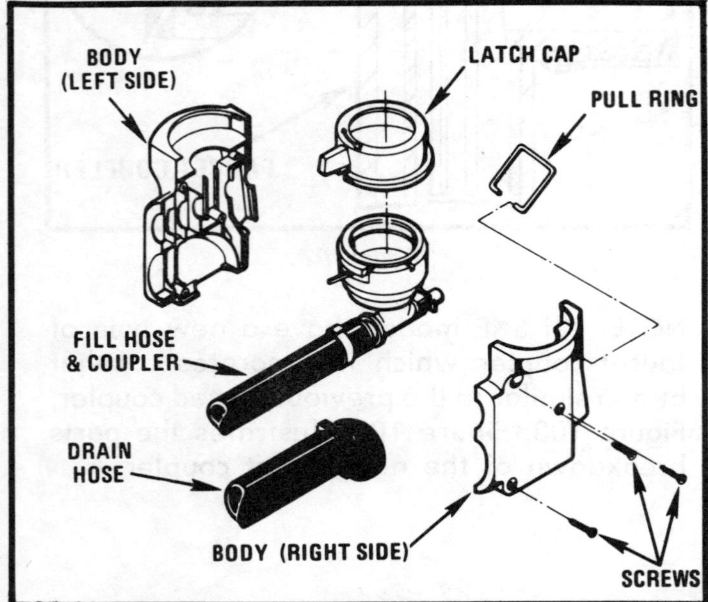

Figure 104

3. Feed the hose through the housing and down the side of the dishwasher.

4. Adjust the hoses in the hose bracket to provide a 31″ length when hoses are fully extended, Figure 107.

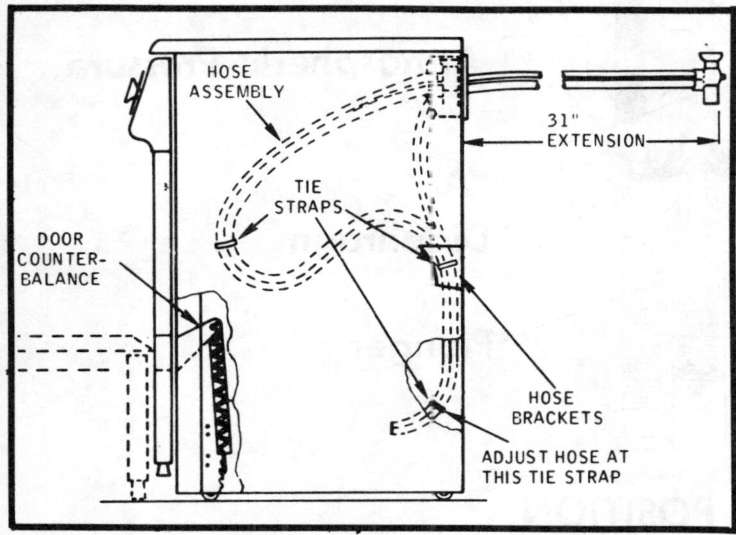

Figure 107

5. To make this adjustment it will be necessary to loosen the tie strap that secures the hoses to the hose brackets. Pull the hoses out to the required length, then secure the hoses to the bracket with the tie strap. The top hose (fill hose) must

be slightly longer than the drain hose to insure proper operation in storage of the hose assembly. Tie straps holding both hoses together must be properly located to maintain the proper length relationship. Replace the dishwasher top.

WATER VALVE, Testing and Removal — All Models.

NOTE: If a condition exists that water enters the tub intermittently at unscheduled times, it may be due to mechanical failure of the valve. Disconnect the service cord from the dishwasher; if water comes into the tub without electric, then the valve has failed mechanically. Replacing the coil will not solve the problem. The valve will have to be cleaned or replaced, see page *61* "Cleaning the Water Valve."

a. Remove wires from solenoid terminals; test with an ohmmeter or continuity tester.

b. If a condition exists that water does not come into the tub, test for voltage at the solenoid coil terminals. If voltage is present and coil does not energize, coil must be replaced.

OPERATION

Figure 108 shows a cross section view of the water inlet valve in its closed and open position. The water pressure at B and C has equalized because the bleed hole opening permits water to flow from B to C. Pressure at A is atmospheric.

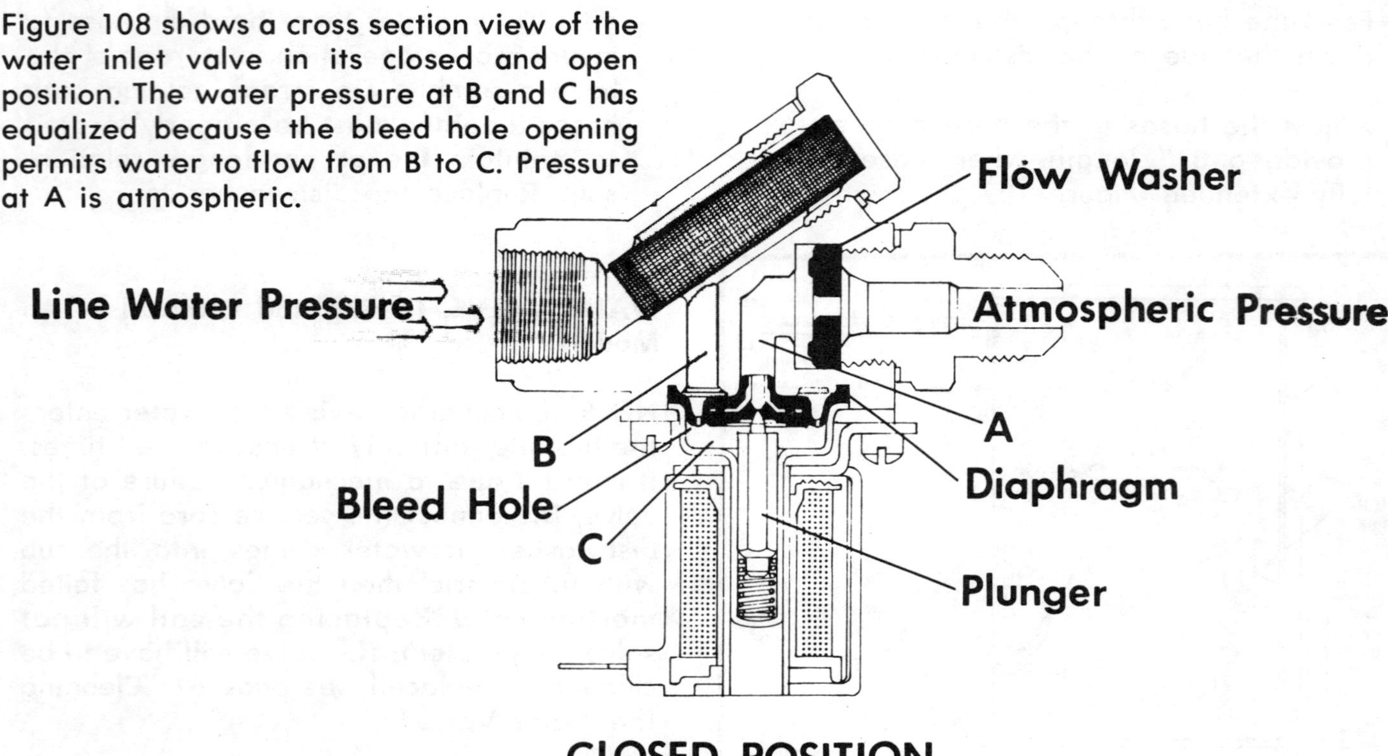

CLOSED POSITION

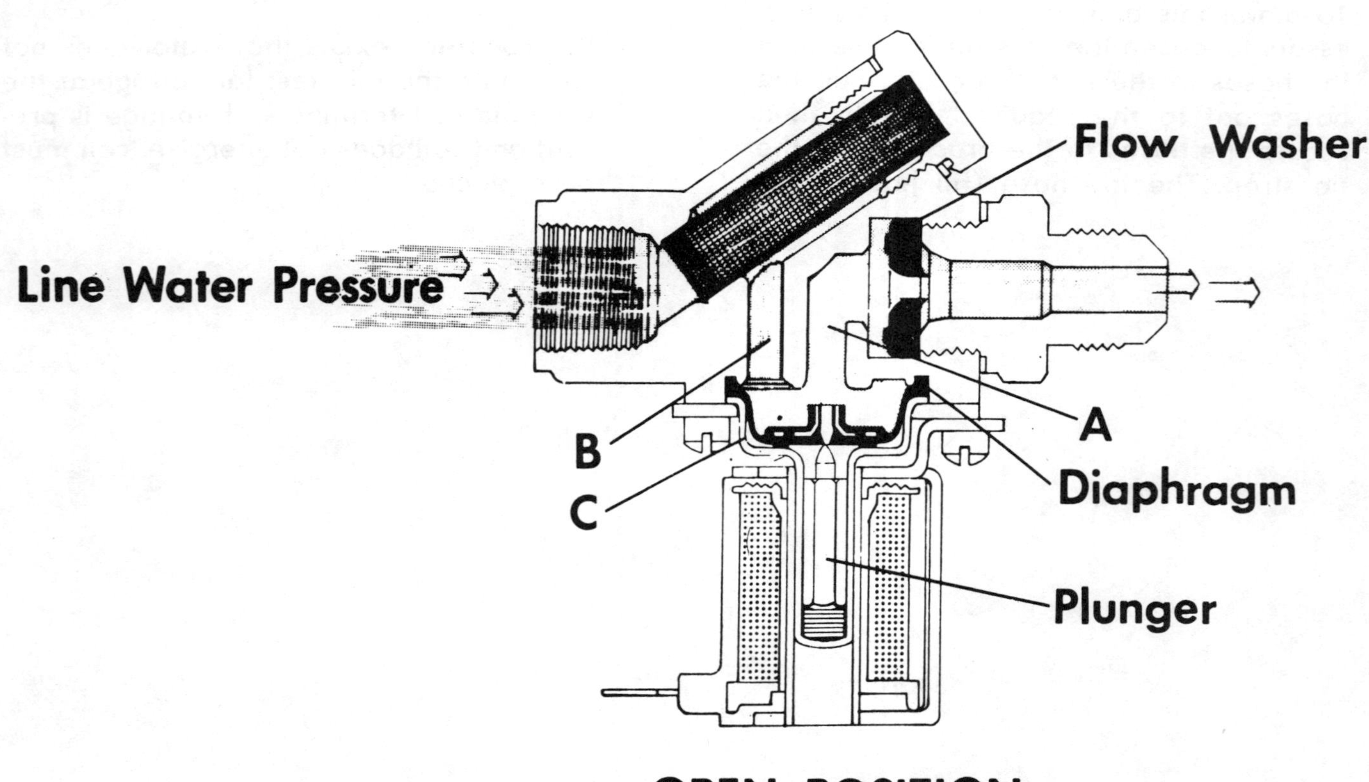

OPEN POSITION

Figure 108

CLEANING THE WATER VALVE

NOTE: A good cleaning agent to use on the internal parts of the valve is a 50% solution of soldering acid (can be purchased at a local hardware store) and water. Mix solution in a glass bowl.

a. Remove valve from valve mounting bracket.

b. Remove wiring from solenoid coil.

c. Remove fill hose.

d. Remove coil from valve. Metal encased coil is a three screw mount, the encapsulated type is one screw mount, but the other two screws must be removed to open the valve. The old style coil has just a retainer at the top; note which side of this latter coil is outside or away from the valve body. It must be reinstalled in the same position as when removed.

NOTE: The diaphragm has an orifice hole and a bleeder hole, Figure 109.

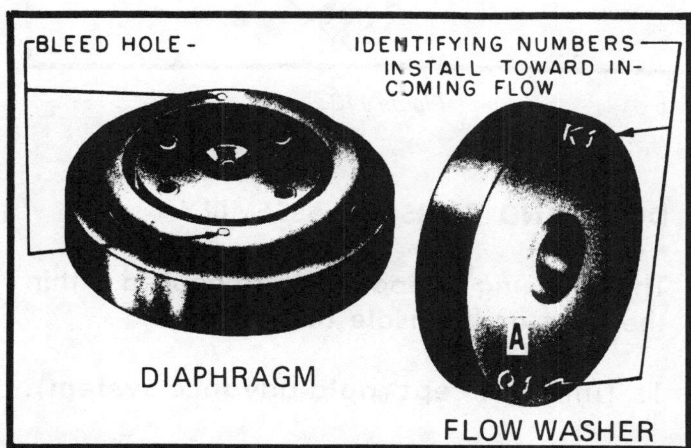

Figure 109

CAUTION: Avoid getting acid on your hands.

Remove the three screws or the four screws according to type of valve.

e. Remove brass cylinder and diaphragm.

f. Inspect valve seat for any nicks or worn spots.

g. Remove spring and plunger from cylinder, Figure 110.

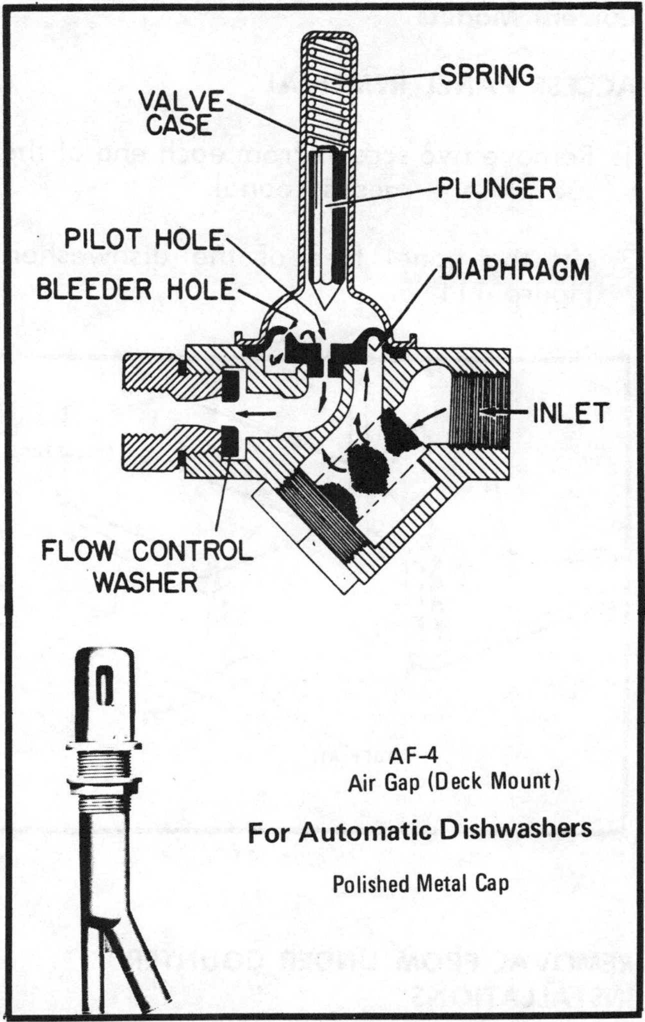

Figure 110

h. A valve kit can be purchased. The kit contains a cylinder, plunger, spring and a diaphragm.

i. If a new kit is not readily available, dip the above parts in the solution as outlined above.

NOTE: When parts are clean of lime deposits, remove and wash with clean water.

j. Reassemble to valve. Replace solenoid coil and install on machine.

UNDER COUNTER AND FRONT LOADING MODELS

WHIRLPOOL DISHWASHERS, Late Thru Current Models.

ACCESS PANEL REMOVAL

1. Remove two screws from each end of the toe plate — access panel.

2. Lift the panel free of the dishwasher, Figure 111.

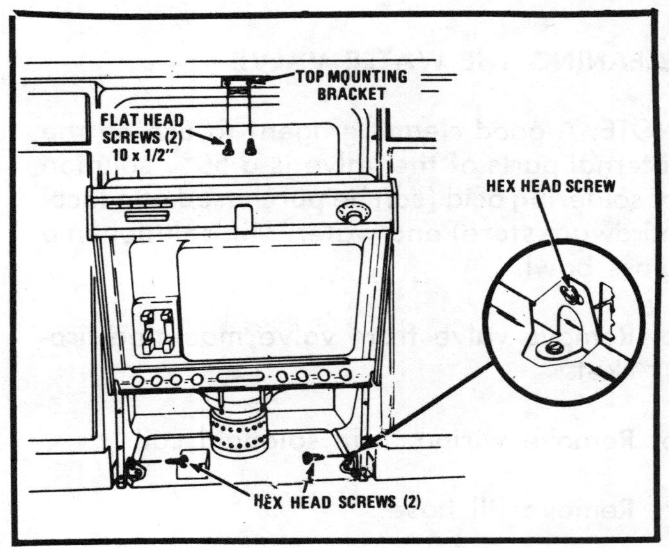

Figure 112

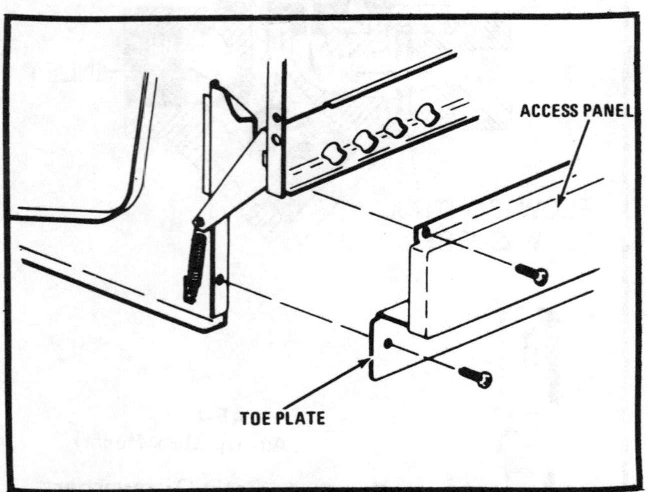

Figure 111

REMOVAL FROM UNDER COUNTER INSTALLATIONS

1. Remove the access panel.

2. Remove the base plate to the dishwasher screws, Figure 112.

3. Remove the top bracket to the counter top screws, Figure 112.

4. Disconnect plug connectors and wiring, Figure 113.

5. Disconnect the drain hose.

6. Disconnect the water inlet.

7. Slide the unit forward and out of recess; base plate will remain.

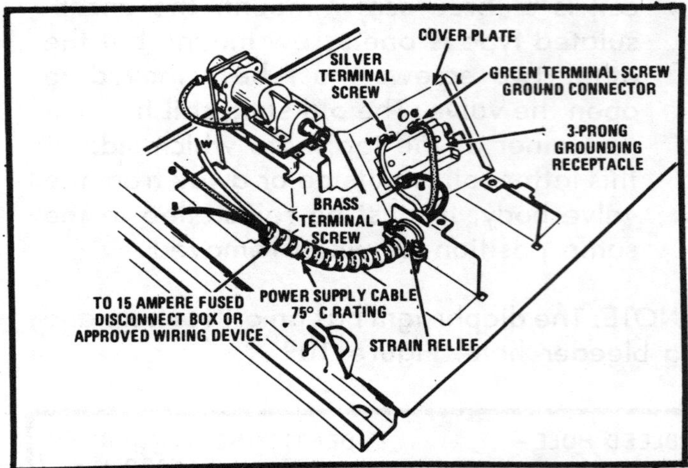

Figure 113

DOOR AND CONSOLE ASSEMBLY

The following components are housed within the door and console assembly:

1. Timer (except rapid advance system).

2. Detergent dispenser.

3. Wetting agent dispenser.

4. Push button selector switch.

5. Wiring harness.

6. Door latch mechanism.

7. Door switch.

8. Pilot light.

9. Vent assembly.

DISASSEMBLY PROCEDURES, Descending Access Panel.

1. Disconnect the power source.

2. Open the door, and remove the rubber push nut that secures the panel springs to the end caps, see Figure 114.

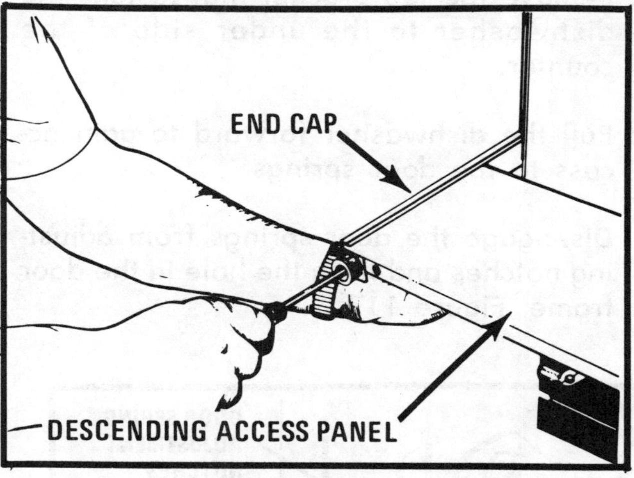

Figure 115

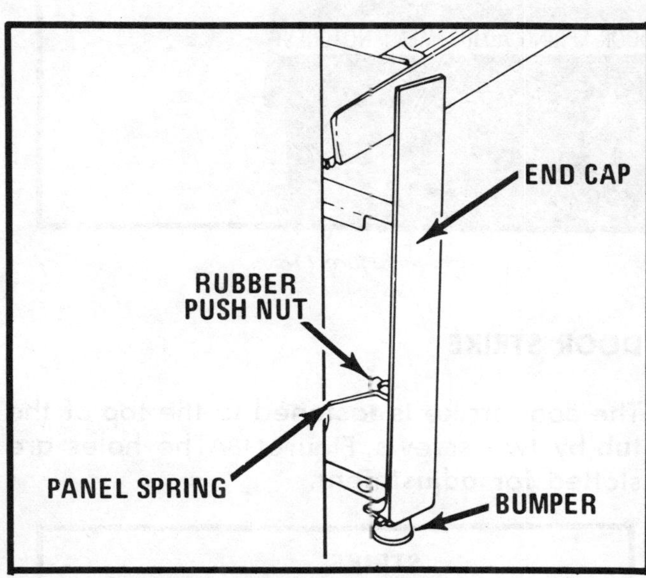

Figure 114

DOOR SPRING, Replacement.

The door springs are interchangeable between all models.

1. Remove the access panel mounting screws, two screws on each side, Figure 111.

2. Disconnect the power source.

3. On undercounter models, remove the two screws from the hold down brackets, Figure 116.

3. Separate the panel springs from the end caps.

4. Close the door. Lift the panel to a horizontal position and loosen the screw holding one of the bumpers to an end cap, Figure 115.

5. Disengage boss on top edge of the end cap from the panel. Pull out on the end cap and remove the assembly from the outer door frame.

6. To reinstall, reverse the procedure.

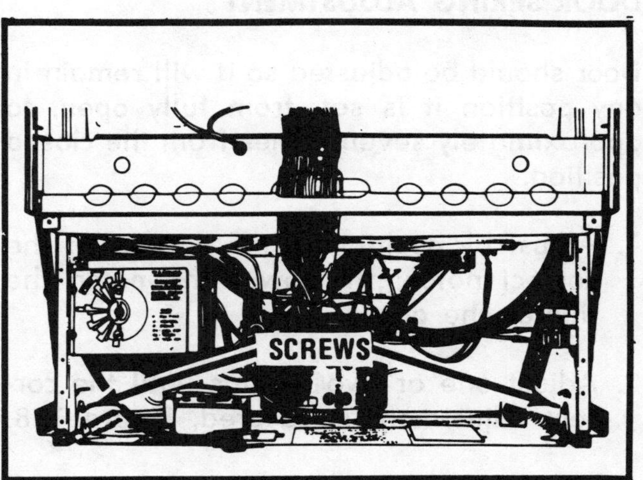

Figure 116

4. Remove the top bracket that fastens the dishwasher to the under side of the counter.

5. Pull the dishwasher forward to gain access to the door springs.

6. Disengage the door springs from adjusting notches and from the hole in the door frame, Figure 117.

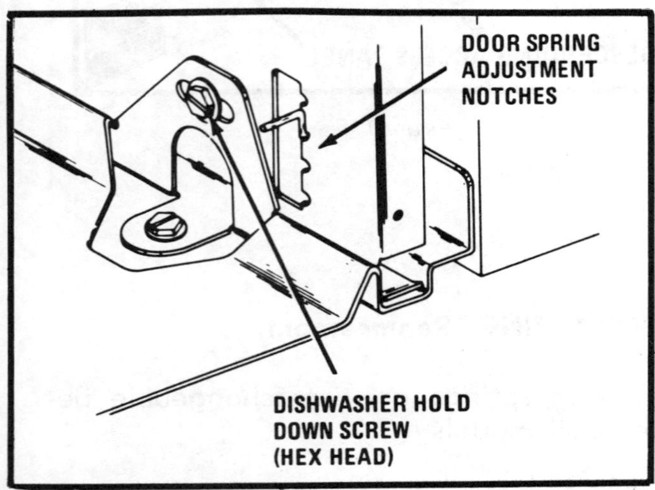

Figure 117

7. The door springs and nylon bushings should be replaced in pairs.

8. To reassemble, reverse procedure.

DOOR SPRING ADJUSTMENT

Door should be adjusted so it will remain in any position it is set, from fully open, to approximately seven inches from the closed position.

1. Adjust by positioning the springs in the correct notch; the lower the notch the tighter the door.

2. Adjust one or two springs until the correct adjustment is secured, Figure 118.

Figure 118

DOOR STRIKE

The door strike is fastened to the top of the tub by two screws, Figure 118A. The holes are slotted for adjustment.

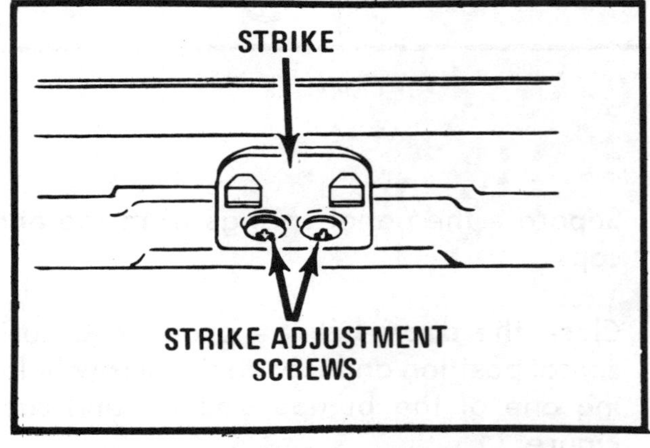

Figure 118 A

1. Loosen the screws to adjust the door strike.

2. On undercounter models, the unit will have to be pulled out from under the counter when replacement of the door strike is necessary.

3. On front loading portables, remove the top assembly to replace the door strike.

4. When adjusting or replacing the door strike, make certain the screws are tightened properly when work is completed.

DISHRACK TRACKS, Upper, Replacement.

1. Remove the upper dishrack from the dishwasher.

2. Extend the track out to line up the holes in the track with the screw slot in roller hubs, Figure 119.

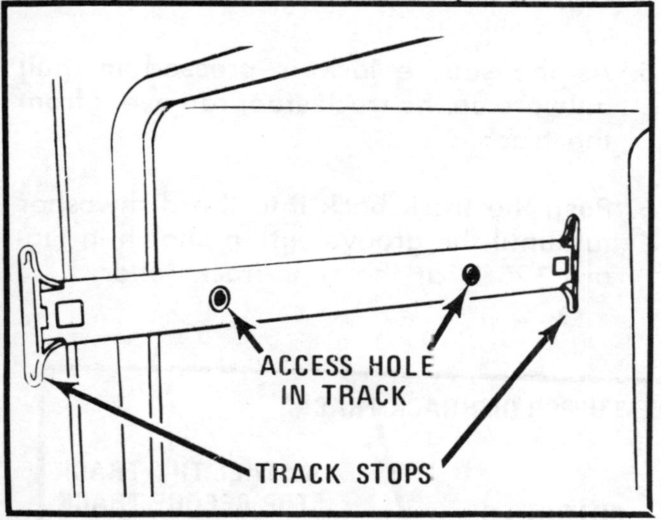

Figure 119

3. Turn twist lock fasteners clockwise only. The locking wire will release from the hole in the tub as you pull out on the wheel assembly.

4. Early models have a twist lock, Figure 120.

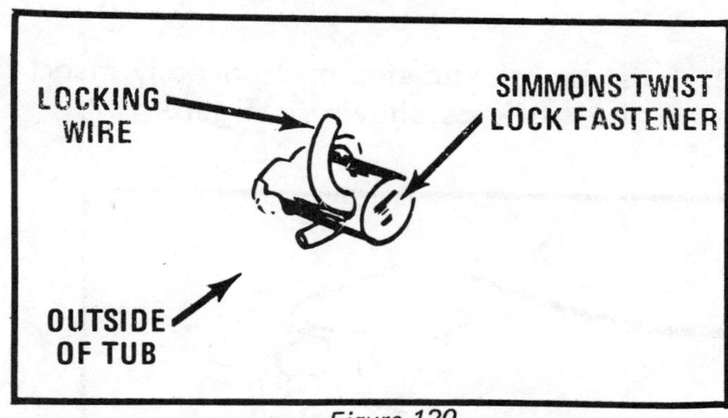

Figure 120

5. Later models have a threaded screw and nut, Figure 121.

6. Reverse procedure to reassemble.

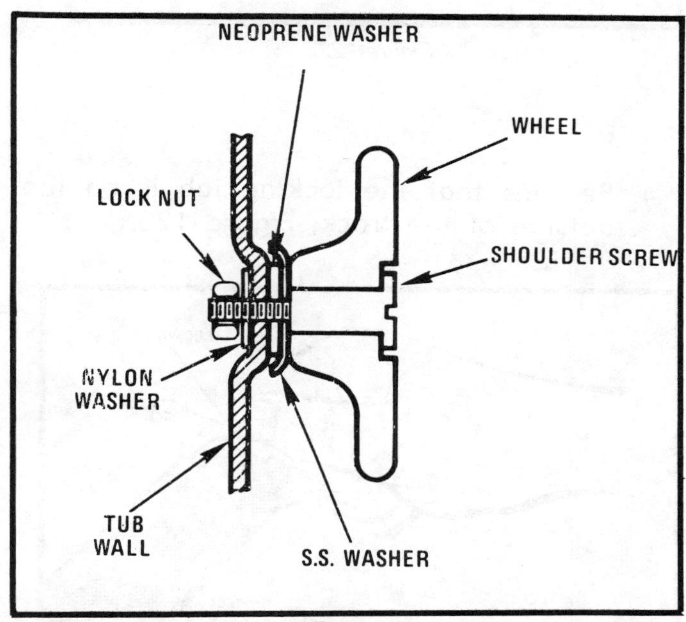

Figure 121

UPPER DISHRACK STOP, Replacement.

1. Remove the track assembly from the dishwasher.

2. Heat the plastic stop in boiling water until it becomes pliable. Insert in the track as shown in Figure 119.

3. Twist the plastic stop and partially insert in the track as shown in Figure 122.

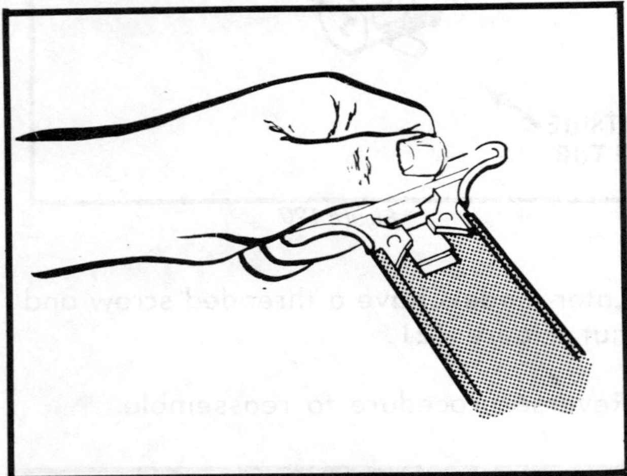

Figure 122

4. Be sure that the locking tab is on the outside of the track, Figure 123.

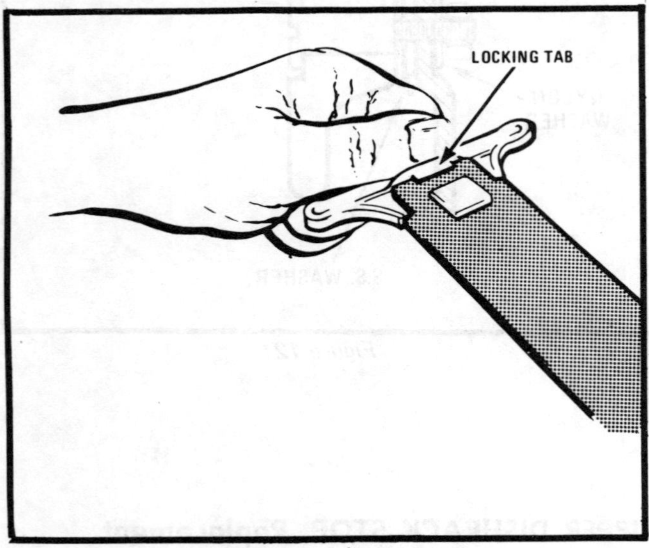

Figure 123

UPPER DISHRACK STOP, Replacement, 1971 Models.

1. Remove the upper dishrack from the dishwasher.

2. Slide the track out to its full length.

3. Insert a small screwdriver blade into the cutout in the track, Figure 124.

4. Pry out the track stop locking square as shown in Figure 124.

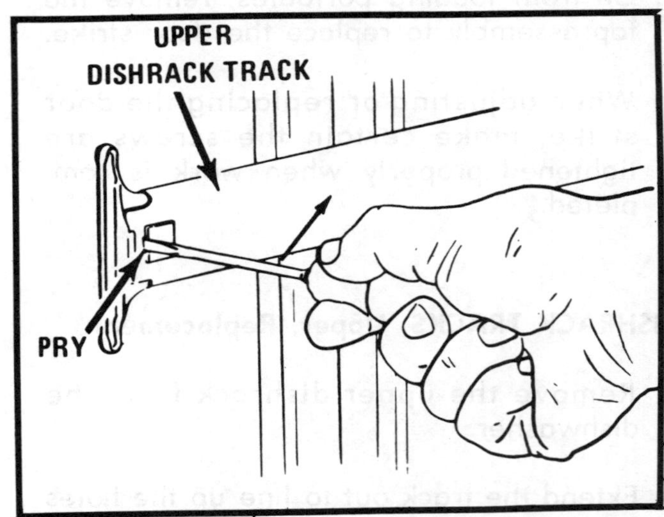

Figure 124

5. As the square lock is pressed in, pull outward on the track stop; remove it from the track.

6. Push the track back into the dishwasher tub until the groove cutout shown in Figure 125 is at the rear track roller.

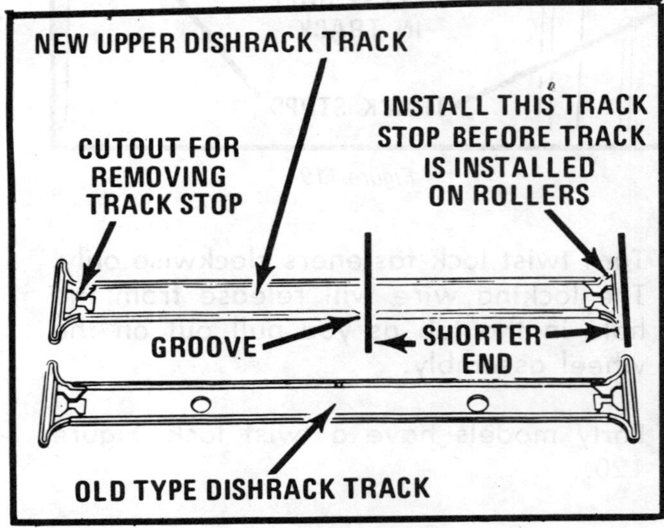

Figure 125

7. Rotate the track in either direction until the track becomes disengaged from the track wheel.

8. The track, track stops or rollers can now be replaced.

9. To reassemble, reverse procedure.

WORK TOP, Replacement.

1. Open the dishwasher door and remove two screws at the top of front trim. Refer to Figure 126.

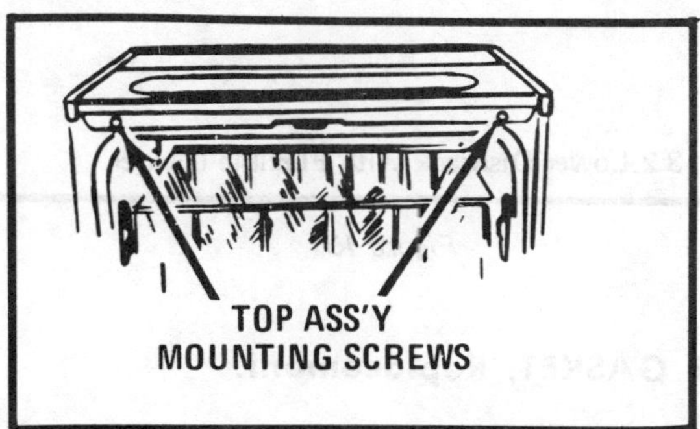

Figure 126

2. Remove the two screws from the rear of the cabinet holding the top assembly to dishwasher, Figure 127.

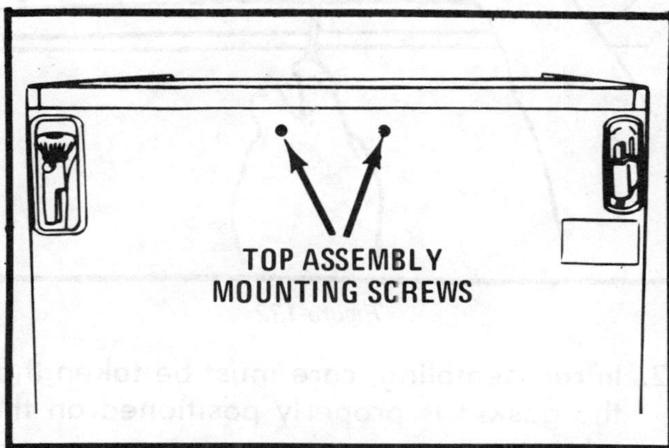

Figure 127

3. Slide the top toward the front of the dishwasher and lift off.

4. On models that have a food warmer at the top, break the connection to the wiring harness by removing the molex connector on top of the dishwasher.

5. To reassemble, reverse procedure.

UPPER DISHRACKS

To remove the upper dishrack, pull the dishrack forward to the stops, and lift up. The wheels will disengage from the track. The wheels can now be replaced.

NOTE: The adjustable type of upper dishracks are removable in the same manner; the following parts are replaceable:

a. Wheels
b. Bearings
c. Linkage
d. Adjusting lever
e. Dishrack lever
f. Dishrack

Disassembly of these parts is accomplished by removing them from the dishrack. When replacing the dishrack lever, the indented crossover portion of the lever must face toward the bottom of the dishrack, seen in Figure 128. Parts identification can be seen in Figure 129.

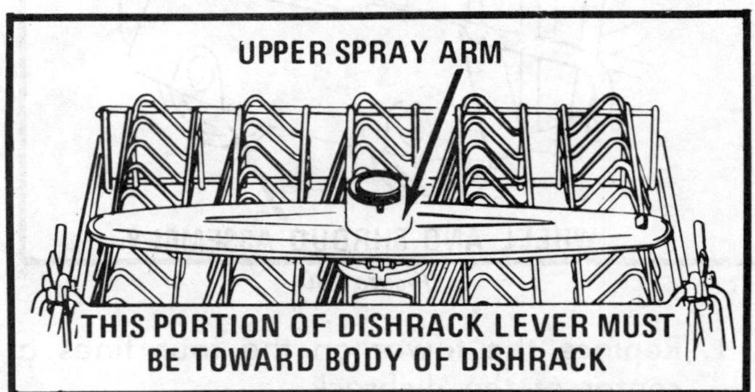

Figure 128

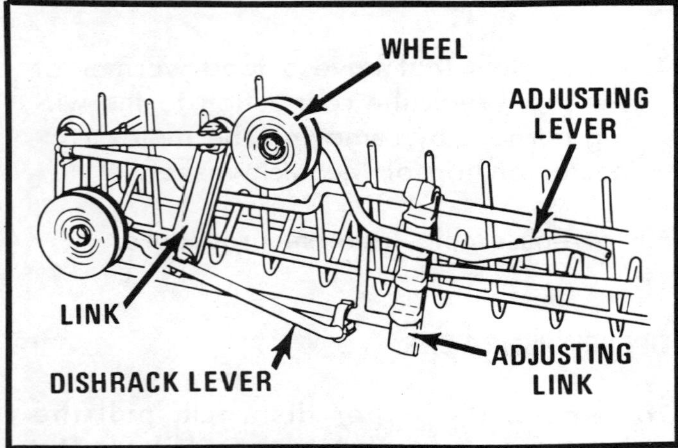

Figure 129

LOWER DISHRACKS, Replacement.

The lower dishrack is supported by eight wheels rolling in the tracks formed into the sides of the tub. The wheel housing assemblies and the plastic tower are replaceable items, Figure 130.

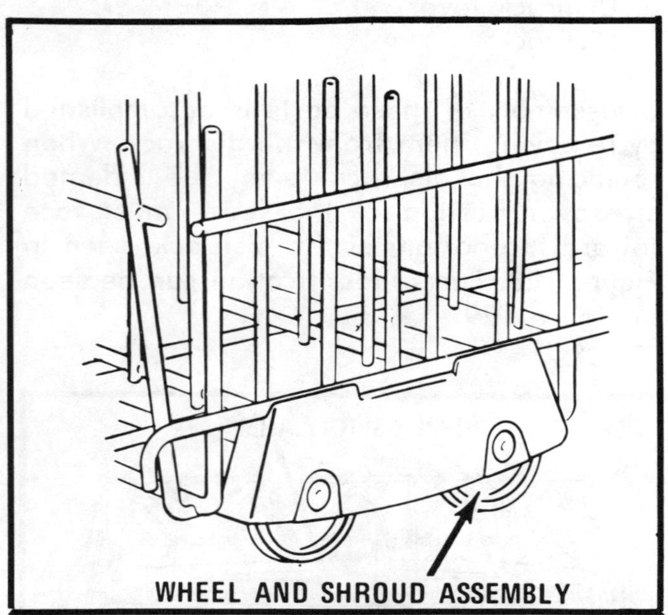

Figure 130

1. Replace the tower on the four tines at center of the dishrack.

2. Gently tap the mounting sockets onto the tines until secure, Figure 131.

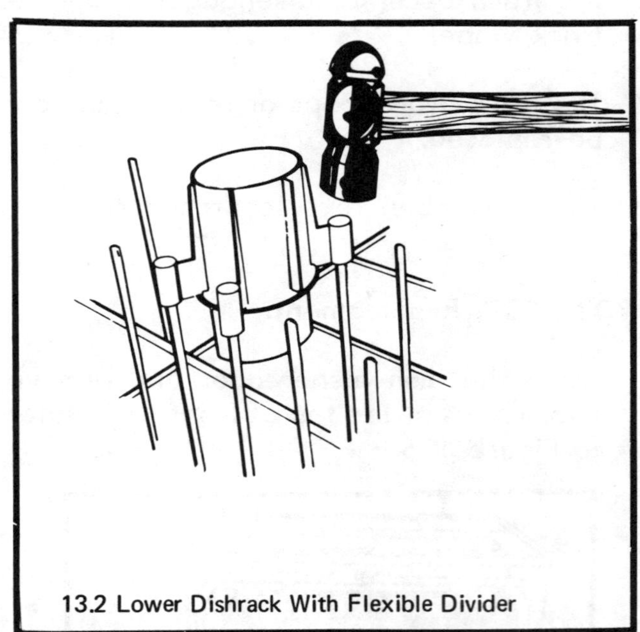

13.2 Lower Dishrack With Flexible Divider

Figure 131

TUB GASKET, Replacement.

1. Remove the six screws from each side of the tub and pull the gasket and retainers free of the tub, see Figure 132.

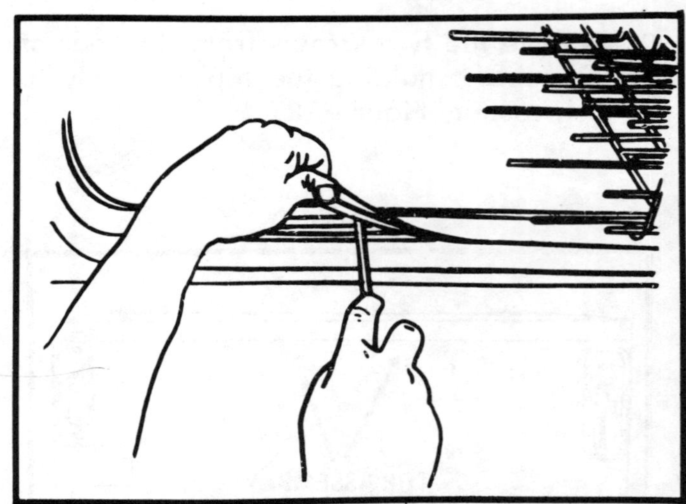

Figure 132

2. In reassembling, care must be taken that the gasket is properly positioned on the tub.

3. Reverse procedure to reinstall gasket.

DISPENSER SOLENOIDS

Early models of the Whirlpool dishwashers used a rinse dispenser that was actuated by a vernay element. Later models use a solenoid to activate the dispensers. A solenoid is used on the detergent dispenser, another solenoid operates the plunger on the rinse additive dispenser. A third solenoid is on the fill valve which is covered in a previous text.

The models FP-40 and FP-50 series of the Whirlpool dishwasher used a detergent cup that was mechanically operated from the timer shaft. With the use of a cam and a spring loaded shaft, the lid of the detergent cup would open at a pre-determined time, see Figure 133. This type of detergent cup

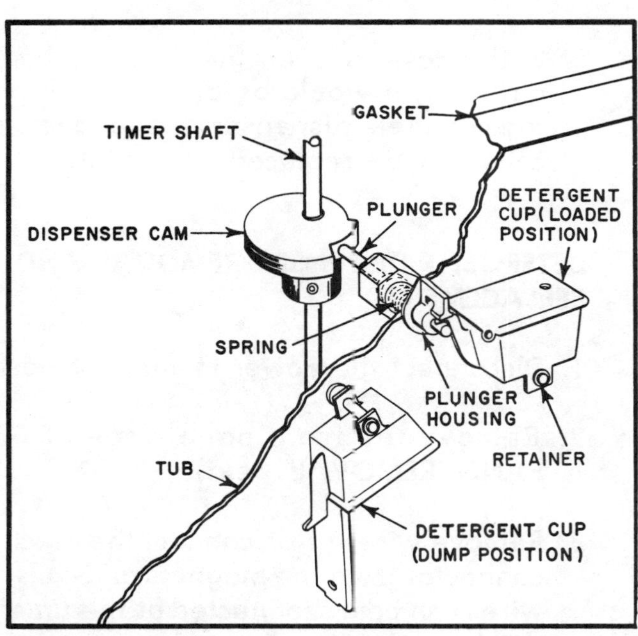

Figure 133

can be checked visually and any broken part can be replaced without too much problem. The main parts to check are 1) The cam, to see if it is tight to the shaft, and in the right place, 2) The pushspring to see if it is rusted and broken; this spring returns the plunger to the cam surface and is important, and 3) The detergent cup itself. These are the three parts that are most likely to be broken.

MECHANICAL DISPENSER PARTS REMOVAL

1. Note the position of the cam in relation to the timer extension shaft. Loosen or remove the set screw; install the new cam on the same spot of the shaft.

2. Back off the plastic nut. This will free the plunger housing and assembly.

3. Remove the detergent cup retainer and slide the detergent cup off of the stud. Reinstall all parts in reverse order.

WETTING AGENT DISPENSER

The wetting agent dispenser is a solenoid operated dispenser that dispenses approximately 2 c.c. of wetting agent during the last rinse cycle only. The wetting agent dispenser can be added to some models that are not so equipped. These kits are available at your Whirlpool distributor, Part number SKK-6.

CHECKING DISPENSER SOLENOID

The wetting agent dispenser solenoid can be tested with a continuity tester; disconnect the power to the dishwasher.

1. Remove the spade connectors to the solenoid.

2. Test across the solenoid terminals for continuity.

3. Place one probe on the terminal of the solenoid; the other probe must be placed on a metal part of the cabinet. If continuity exists, it indicates the solenoid coil is burned to ground and must be replaced.

WETTING AGENT DISPENSER REMOVAL

1. Before removal of the wetting agent dispenser, test the pump and solenoid with a direct line to the solenoid.

2. Disconnect the power to the dishwasher.

3. Remove the front panel from the dishwasher, (see "FRONT PANEL REMOVAL" text).

4. Remove the leads from the solenoid coil.

5. Remove the dispenser caps and nuts on the inside door panel, and remove the dispenser from the holes.

6. To reassemble reverse procedures.

DETERGENT DISPENSER

Most models of the Whirlpool dishwasher have an automatic dispenser that is solenoid operated. The detergent dispenser has been revised several times since its original concept and a new detergent dispenser kit is available. The new kit may not be the same as the one that is removed; however, it will do the same work. Complete installation instructions are supplied with the kit; some wiring harness changes are required for this new installation.

DETERGENT DISPENSER SOLENOID TESTING, Figure 134.

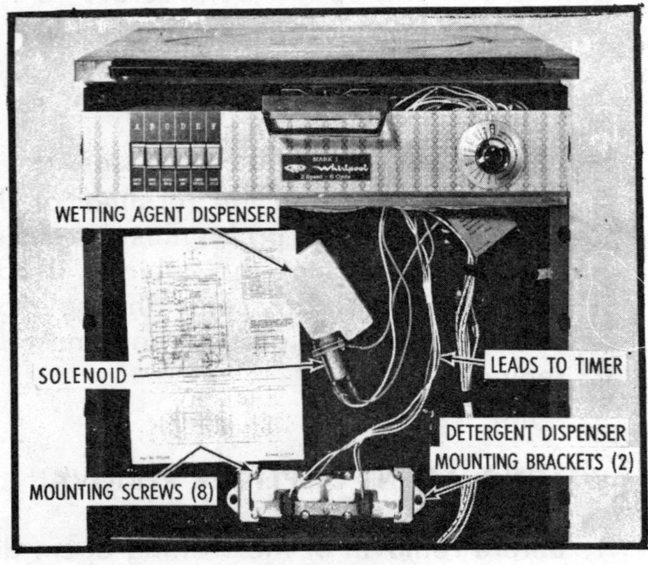

Figure 134

1. Disconnect the power to the dishwasher.

2. Remove the front panel, (see "FRONT PANEL REMOVAL" test).

3. Check each coil individually with a continuity tester. The leads must be removed from the coils at this time. If the detergent dispenser has an electric connector, it must be disconnected and checked from the prongs of the connector.

4. Close the dispenser cover and check the coils with a direct line. Care must be taken when connecting a direct line to the prongs of the connector. It would be well to make up some short leads with the proper connectors placed on the ends of the leads to connect to the electrical connector.

5. If the test fails or there is mechanical difficulty, it would be prudent to replace the complete dispenser; the magnets and coils can be replaced separately.

DETERGENT DISPENSER REMOVAL AND REPLACEMENT

1. Disconnect the power to the dishwasher.

2. Remove the front panel (see "FRONT PANEL REMOVAL" text).

3. Remove wires or disconnect the electrical connector from the magnets and coils; the wires can be disconnected at the timer, or the harness disconnect.

4. Remove the eight screws that secure the dispenser to the inside door panel. On later models two brackets secure the dispenser to the door panel.

5. Open the cabinet door and remove the dispenser assembly.

6. For reassembly reverse procedures.

CONVERSION KIT 569056 FOR FRONT LOADING PORTABLE DISHWASHERS

The following parts are required to make an approved built-in installation of a front loading portable dishwasher.

Quantity	Part No.	Part Description
3	718702	Grommets
1	718205	Terminal Box
1	714500	Terminal Box Cover
3	717232	Screw
1	718178	Screw (Ground)
1	717664	Drain Hose (Drain Loop)
2	712138	Hose Clamp
1	717563	Strain Relief
1	16123	Washer
1	596490	Hose Strap
1	718206	Conversion Instructions

FIELD TECHNICIAN'S INSTRUCTIONS FOR CHANGING A WHIRLPOOL FRONT LOADING PORTABLE DISHWASHER INTO A BUILT-IN MODEL

The following is the procedure for the conversion of a front loading portable model dishwasher into a built-in model. This procedure consists mainly of installing service lines (power supply, water supply and drain outlet) and removing several parts from the existing dishwasher assembly.

PRE-INSTALLATION REQUIREMENTS

Provide proper cabinet opening, Figure 135.

1. Width — 24¼".

    a. The opening must be square and have no taper.

2. Height — 34½" minimum.

    a. This dimension should be measured from the finished floor surface.
    b. If floor covering is added after the dishwasher is installed, it may interfere with removal if this ever becomes necessary.

3. Depth — 24" minimum.

    a. Cabinets of lesser depth will result in excessive protrusion of the dishwasher.

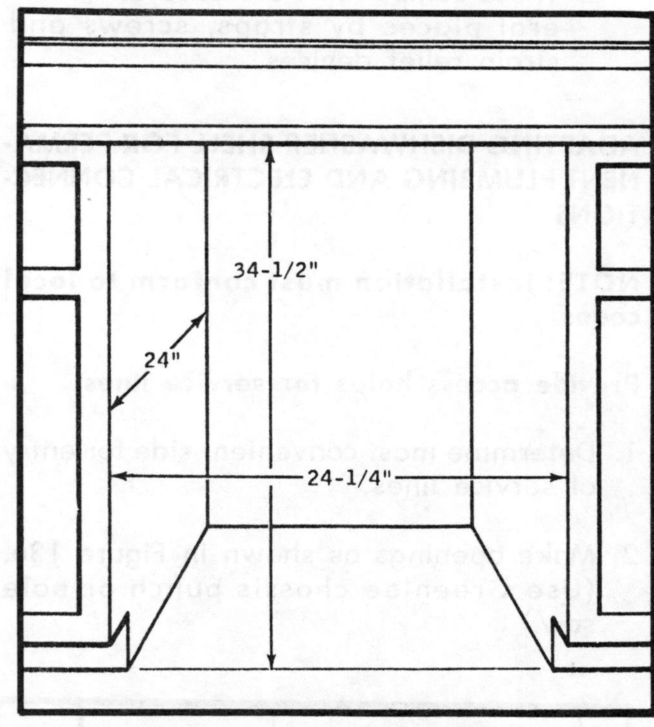

Figure 135

Adapt dishwasher for installation.

1. Remove top from dishwasher.

    a. Open door of dishwasher and remove two screws securing top front trim to top tub flange.
    b. Remove two screws from rear top center of dishwasher cabinet.
    c. Lift top free of cabinet.

2. Remove casters from bottom of cabinet.

3. Remove two bumpers from bottom of descending access panel and replace screws.

4. Remove power cord, water hoses and housing assemblies (also cord reels on applicable models).

   a. Disconnect the power cord from the wiring harness by clipping the wires on the power cord at the strain relief. The black and white leads of the wiring harness will be run into the junction box that mounts to the left leg.
   b. These components are secured in several places by straps, screws and strain relief devices.

ADAPTING DISHWASHER SHELL FOR PERMANENT PLUMBING AND ELECTRICAL CONNECTIONS

NOTE: Installation must conform to local codes.

Provide access holes for service lines.

1. Determine most convenient side for entry of service lines.

2. Make openings as shown in Figure 136. (Use Greenlee chassis punch or hole saw.)

3. Position dishwasher as desired in kitchen cabinets.

4. Secure to floor with screws through holes in bottom flanges of dishwasher cabinet.

5. Install the three grommets in the access holes for protection of the service lines.

Drain Loop Installation

1. Install the drain loop furnished in the kit between the dishwasher cabinet and tub.

2. Use the strap provided to secure the top of the loop to the flange of the cabinet. See Figure 137.

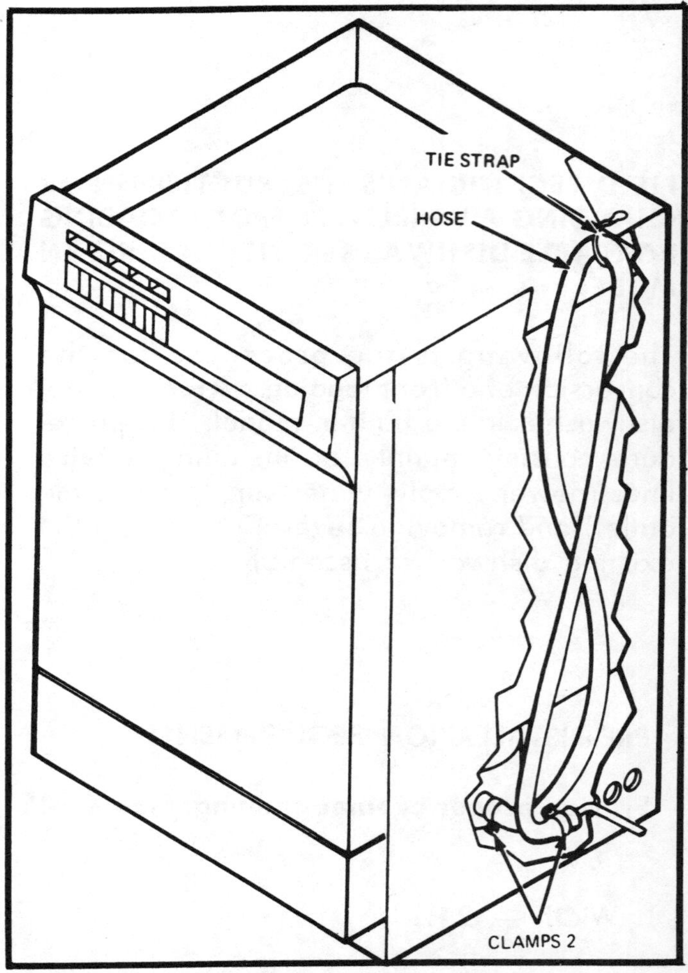

Figure 137

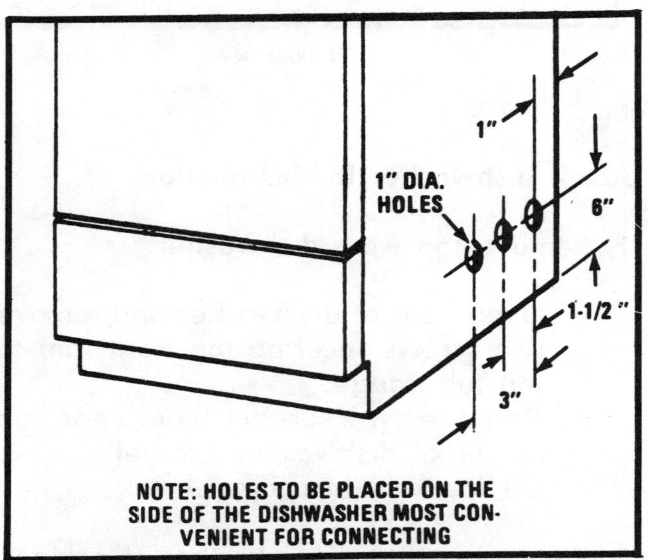

Figure 136

Install Dishwasher in Cabinet Opening

1. Position dishwasher as desired in kitchen cabinets.

2. Secure to floor with screws through holes in bottom flanges of dishwasher cabinet.

Install Electrical Junction Box

1. Mount the junction box to the left side of the dishwasher. See Figure 138.

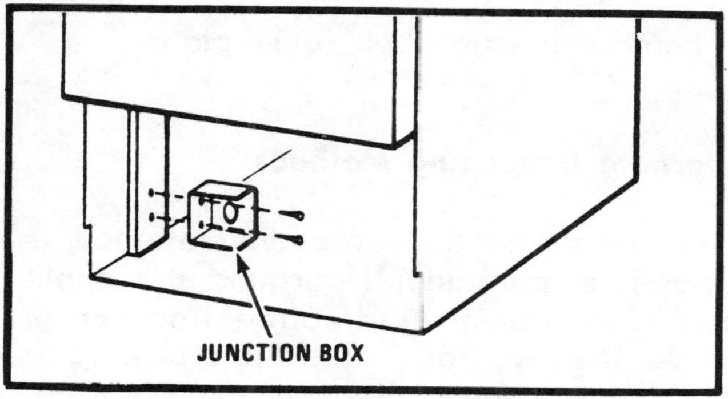

Figure 138

2. Push the black and white leads of the wiring harness through the strain relief about 4 inches.

3. Install the strain relief in the hole provided in the top of the junction box.

Water Connection

1. Remove the inlet water hose from the faucet coupler.

2. Connect the inlet water hose to the hot water supply line--the connector is a 1/8" IPS (iron pipe size) male fitting.

3. Reconnect the other end of the water inlet hose to the fill valve.

4. See Figure 139, for rough-in locations.

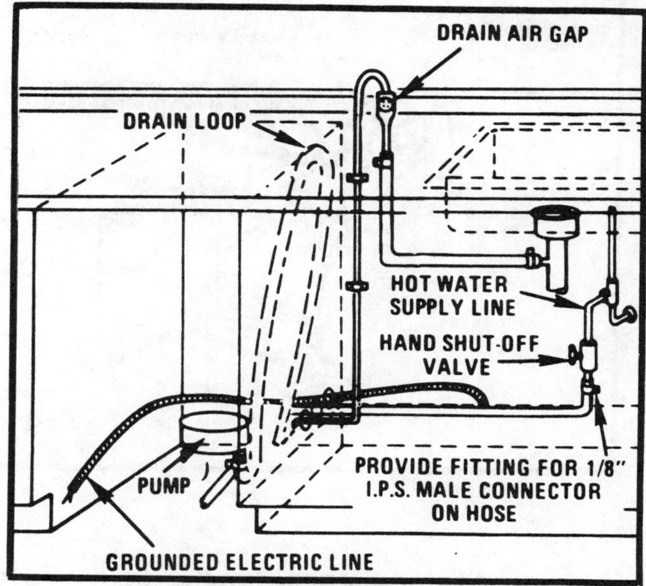

Figure 139

Connect Drain Line

1. Provide a 5/8" O.D. copper drain line and extend the drain line through one of the access holes in the cabinet as shown in Figure 139.

2. Connect one end of the drain loop to the outlet of the dishwasher pump.

3. Connect outlet end of rubber dishwasher drain loop to the copper tubing with hose clamp supplied in kit. The recommended method of connecting the dishwasher to the house plumbing is with a drain air gap to eliminate back siphoning.

ELECTRICAL INSTALLATION

Observe All Governing Codes and Ordinances

A 120 volt 60 cycle AC only 15 ampere fused electrical supply is required (time-delay fuse or circuit breaker is recommended). It is recommended that a separate circuit serving only this appliance be provided. Do not use an extension cord, Figure 140.

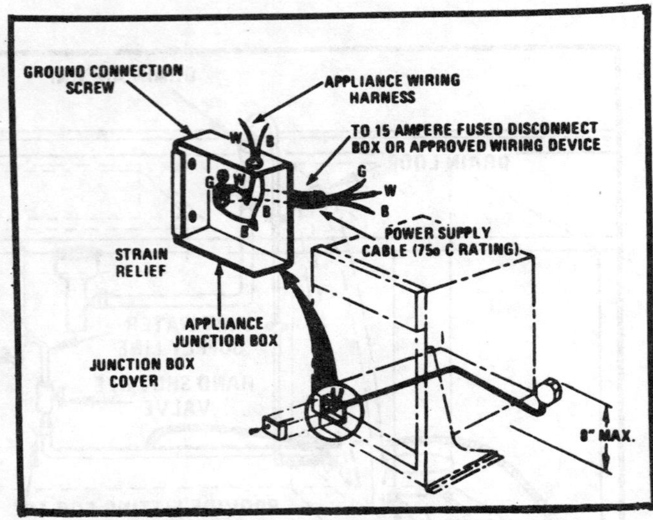

Figure 140

Connect the appliance directly to the fused disconnect (or circuit breaker) box or a junction box through flexible armored or non-metallic sheathed cable (with ground wire). Wire must be suitable for 75° C. A suitable strain relief must be provided at each end of the power supply cable (at appliance and at the junction box).

Electrical Connection

Electrical ground is required on this appliance.

Recommended Electrical Connection and Grounding Method

This appliance must be permanently grounded in accordance with the National Electrical Code and local codes and ordinances.

a. Remove cover plate from appliance junction box located behind the toe plate access panel.

b. Connect the white and black wires of the power supply cable to the white and black leads in the junction box. See adjoining figure.

c. Connect the green ground wire of the power cable to the ground connection screw located inside the junction box. See adjoining figure.

d. Replace junction box cover plate.

Alternate Grounding Methods

If the recommended grounding method is impossible, permanently ground the appliance from the ground connection screw inside the appliance junction box to a grounded cold water pipe* using a separate, green colored insulated conductor of appropriate size (#18 minimum). THIS HOWEVER, IS NOT RECOMMENDED. Do not ground to a gas pipe. Do not connect to electrical supply until appliance is permanently grounded.

*Cold water pipe must have metal continuity to electrical ground and not be interrupted by plastic, rubber or other electrically insulating connectors (including water meter or pump) without adding a jumper wire at these connections.

SECTION 4

PARTS LISTS

The following parts lists are representative of the majority of the most popular parts used in servicing Whirlpool Dishwashers. Mainly, they are shown as an aid in assembly sequence and to show the nomenclature of the various parts.

When ordering parts always give the full model and serial number of the dishwasher. These numbers are found on a metal identification plate.

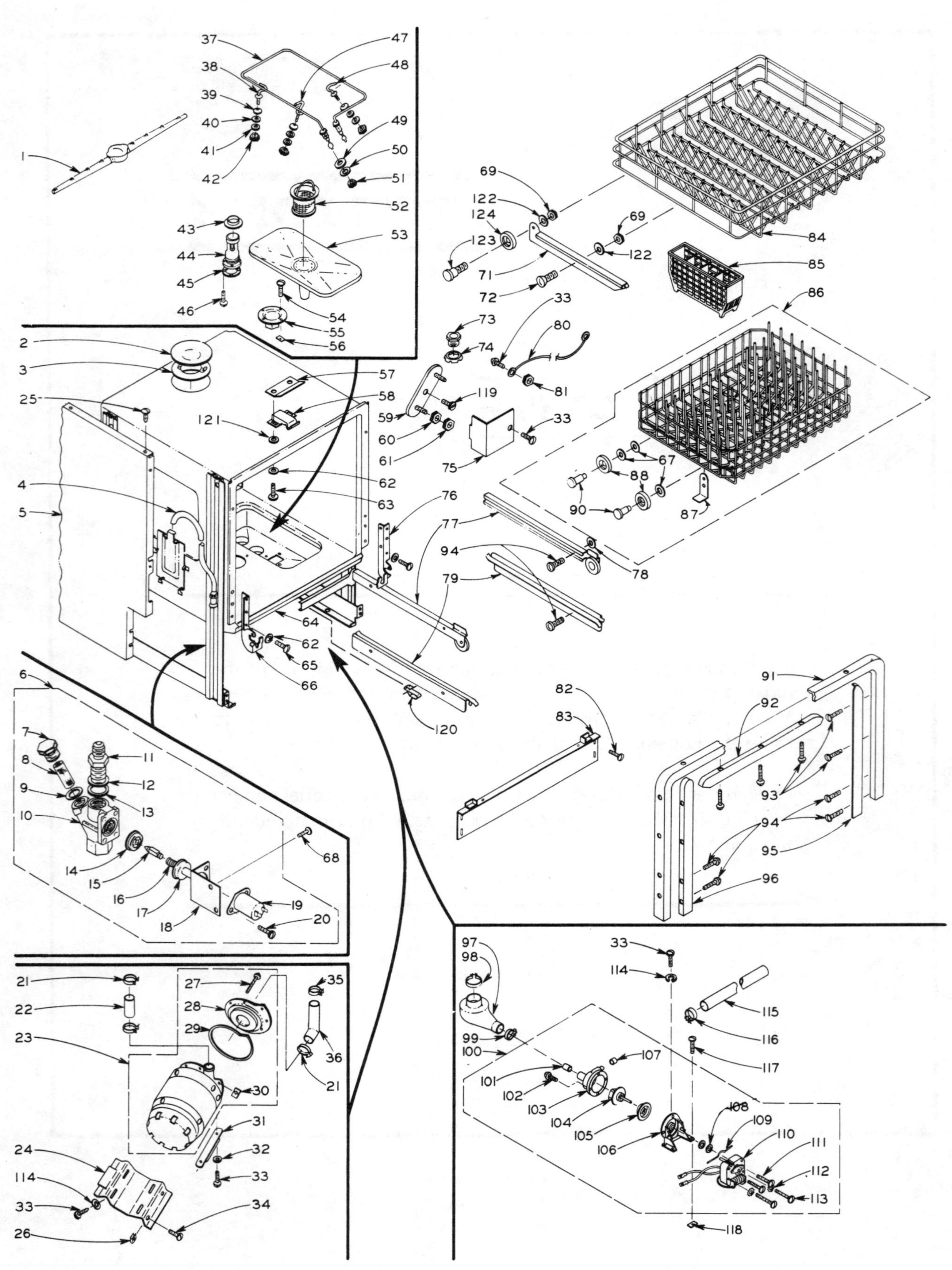

TYPICAL OF UNDERCOUNTER DISHWASHERS
Order Parts by Model Number

Illus. No.	DESCRIPTION	Illus. No.	DESCRIPTION	Illus. No.	DESCRIPTION
1	Swirl Arm	39	Rubber Washer	88	Wheel
2	Relief Valve	40	Washer Retainer	90	Rivet
3	Baffle	41	Fibre Washer	91	Gasket
4	Water Inlet Tube	42	No. 10-32 Nut	92	Retainer – Top
5	Side Panel – LH	43	Stationary Hub Bearing	93	No. 10-32 x 1/2 Screw
	White	44	Stationary Hub	94	No. 10-32 x 3/8 Screw
	Pink	45	Hub Gasket	95	Retainer – RH
	Yellow	46	No. 8-32 x 7/16 Screw	96	Retainer – LH
	Copper	47	Heater Support	97	Drain Pump Boot
	Turquoise	48	Heater Support	98	Clamp
	Side Panel – RH	49	Fibre Washer	99	Clamp
	White	50	1/2 Washer	100	Drain Pump Assembly
	Pink	51	7/16-20 Nut	101	Inlet Insert
	Yellow	52	Sump Cup	102	No. 6-32 x 3/8 Screw
	Copper	53	Sump Screen	103	Pump Body
	Turquoise	54	No. 8-32 x 9/16 Screw	104	Impeller
6	Water Inlet Valve Assembly	55	Pressure Switch	105	Shaft Liner
7	Screw Plug	56	Speed Nut	106	Mounting Bracket
8	Screen	57	Strike Plate	107	Discharge Insert
9	Gasket	58	Latch Strike	108	Slinger Washer
10	Valve Body	59	Terminal Block	109	Grounding Clip
11	Outlet Fitting	60	No. 10-32 Nut	110	Motor
12	Gasket	61	No. 10-32 Nut	111	No. 6-32 x 2 1/2 Screw
13	Flow Washer	62	Fibre Washer	112	No. 6 Lockwasher
14	Diaphragm	63	No. 8-32 x 1/2 Screw	113	No. 6-32 x 1 3/4 Screw
15	Armature	64	Tub	114	Lockwasher
16	Armature Spring	65	No. 8-32 x 7/16 Screw	115	Drain Hose (SJU-70-0)
17	Armature Guide	66	Hinge Bracket	116	Clamp (SJU-70-0)
18	Mounting Bracket	67	Washer	117	No. 6-32 x 7/16 Screw
19	Solenoid	68	1/4-20 x 3/8 Screw	118	Speed Nut
20	Screw	69	No. 8-32 Nut	119	No. 8-32 x 3/8 Screw
21	Clamp	71	Slide – RH	120	Speed Clip
22	Pump Tube		Slide – LH	121	Lockwasher
23	Circulating Pump and Motor Assembly	72	No. 8-32 x 5/8 Screw	122	Washer
24	Pump Cradle Support	73	Bushing	123	Shoulder Screw
25	No. 8-15 x 1/2 Screw	74	Nut	124	Wheel
26	Speed Clip	75	Terminal Box Cover		
27	No. 8-18 x 1 Screw	76	Hinge Bracket		
28	Cover	77	Upper Track – RH		
29	Gasket		Upper Track – LH		
30	Speed Nut	78	Washer		
31	Motor Mount	79	Lower Track – RH		
32	Lockwasher		Lower Track – LH		
33	No. 8-32 x 3/8 Screw	80	Ground Wire		
34	No. 10-32 x 1/2 Screw	81	No. 8-32 Nut		
35	Hose Clamp	82	No. 10-32 x 3/8 Screw		
36	Pump Elbow	83	Toe Plate		
37	Heater	84	Upper Dishrack		
38	Heater Support	85	Silverware Basket		
		86	Lower Dishrack Assembly		
		87	Hold Down		

TYPICAL OF UNDERCOUNTER DISHWASHERS
Order Parts by Model Number

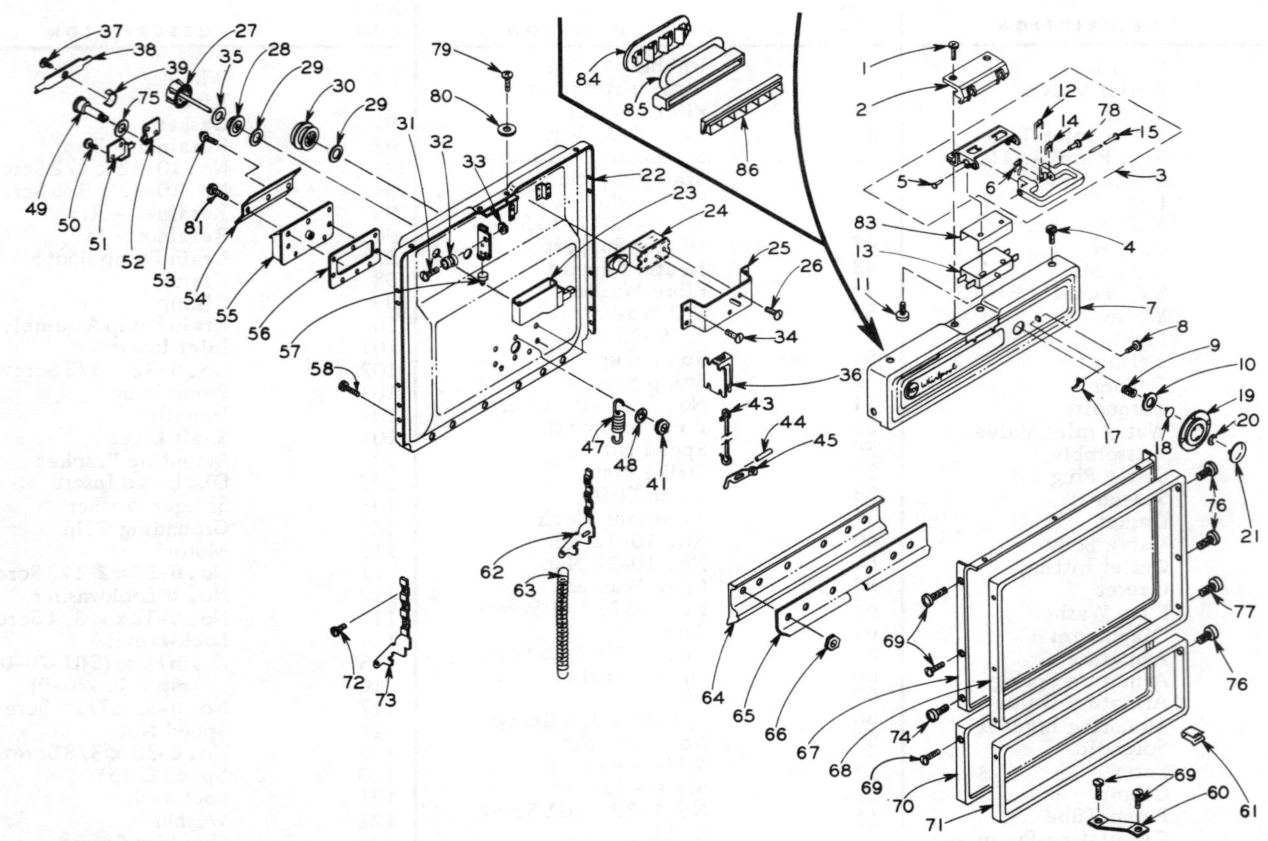

Illus. No.	DESCRIPTION	Illus. No.	DESCRIPTION	Illus. No.	DESCRIPTION
1	No. 8–32 x 1/2 Screw	35	Gasket	68	Door Panel Trim (Used W/Overlay)
2	Handle	36	Dispenser Solenoid		Door Panel Trim (Flush)
3	Latch Assembly	37	No. 6–20 x 5/16 Screw		Door Panel Trim (Used W/Wood Overlay)
4	No. 8–32 x 7/16 Screw	38	Retainer Cover	69	No. 8–15 x 1/2 Screw
5	Pin	39	Cover Spring	70	Access Panel
6	Handle Spring	41	No. 8–32 Nut		White
7	Escutcheon	43	Linkage		Pink
8	No. 8–32 x 3/16 Screw	44	Roll Pin		Yellow
9	Spring	45	Lever Arm		Copper
10	Cam	47	Return Spring		Brushed Chrome
11	No. 6–32 x 7/8 Screw	48	Lockwasher		Turquoise
12	Spring	49	Cam	71	Access Panel Trim (Used W/Overlay)
13	Interlock Switch	50	No. 6–20 x 3/4 Screw		Access Panel Trim (Flush)
14	Spring Bolt	51	Cover		Access Panel Trim (Used W/Wood Overlay)
15	Pin	52	Cover	72	No. 8–32 x 1/2 Screw
17	Finger Stop	53	No. 8–32 x 1/2 Screw	73	Hinge – LH
18	Torque Spring	54	Support Frame	74	No. 8–32 x 1/2 Screw
19	Timer Dial	55	Detergent Cup	75	"O" Ring
20	Retainer	56	Gasket	76	Screw
21	Button	57	Bumper	77	Screw
22	Door	58	No. 8–32 x 7/16 Screw	78	Pin
23	Dispenser	60	Panel Spring	79	No. 8–32 x 1/4 Screw
24	Service Timer (Production Timers 711763 or 711770)	61	Clip	80	Fibre Washer
		62	Hinge – RH	81	Screw
25	Timer Bracket	63	Door Spring	83	Switch Insulation
26	No. 10–32 x 1/4 Screw	64	Door Gasket	84	Grille and Screen
27	Dipstick Cap	65	Gasket Retainer	85	Boot
28	Nut	66	No. 8–32 Nut	86	Grille
29	Grommet	67	Door Panel		
30	Drip Cap		White		
31	No. 10–32 x 5/8 Screw		Pink		
32	Grommet		Yellow		
33	No. 10–32 Nut		Copper		
34	No. 8–32 x 3/8 Screw		Brushed Chrome		
			Turquoise		

TYPICAL OF UNDERCOUNTER DISHWASHERS
Order Parts by Model Number

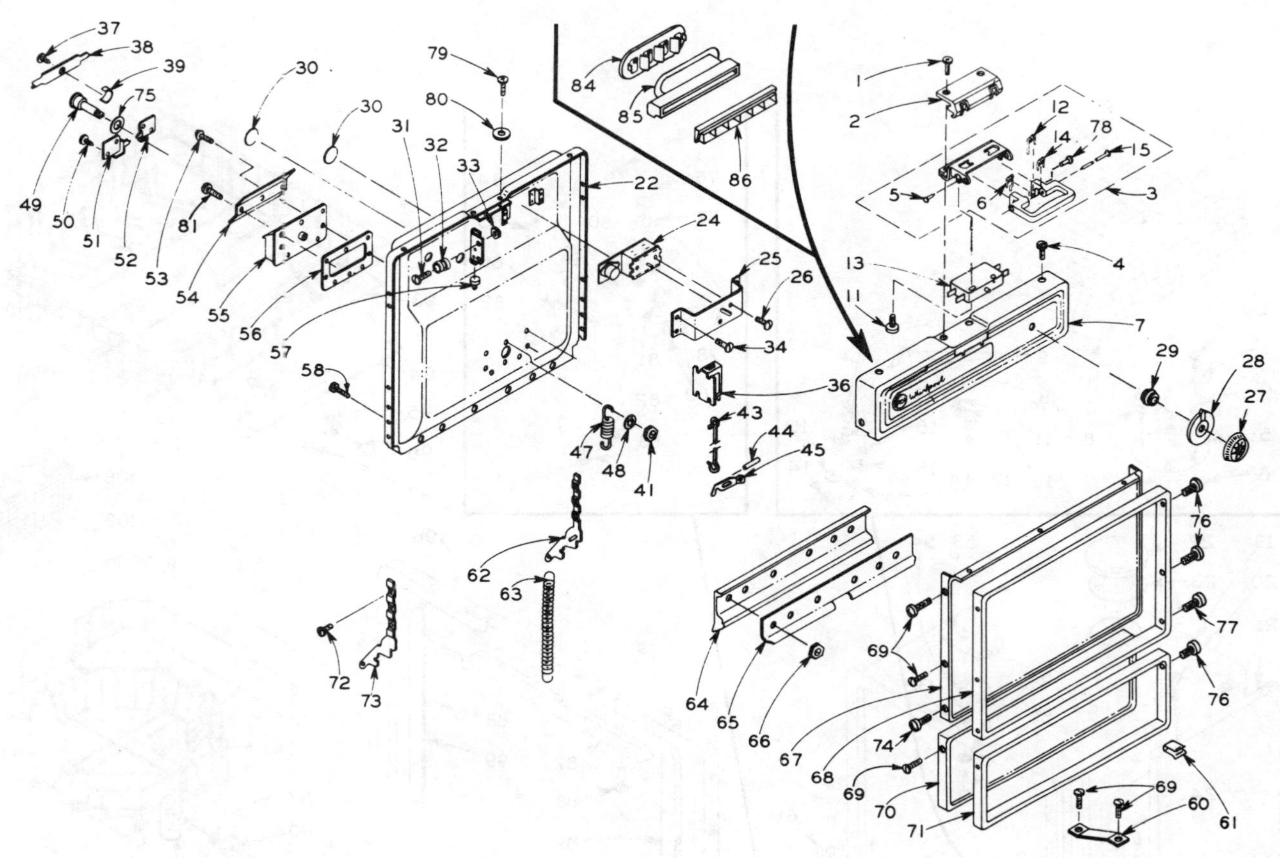

Illus. No.	DESCRIPTION	Illus. No.	DESCRIPTION	Illus. No.	DESCRIPTION
1	No. 8-32 x 1/2 Screw	47	Return Spring	69	No. 8-15 x 1/2 Screw
2	Handle	48	Lockwasher	70	Access Panel
3	Latch Assembly	49	Cam		White
4	No. 8-32 x 7/16 Screw	50	No. 6-20 x 3/4 Screw		Pink
5	Pin	51	Cover		Yellow
6	Handle Spring	52	Cover		Copper
7	Escutcheon	53	No. 8-32 x 1/2 Screw		Brushed Chrome
11	No. 6-32 x 7/8 Screw	54	Support Frame		Turquoise
12	Spring	55	Detergent Cup	71	Access Panel Trim
13	Interlock Switch	56	Gasket		(Used W/Overlay)
14	Spring Bolt	57	Bumper		Access Panel Trim
15	Pin	58	No. 8-32 x 7/16 Screw		(Flush)
22	Door	60	Panel Spring		Access Panel Trim
24	Timer	61	Clip		(Used W/Wood Overlay)
25	Timer Bracket	62	Hinge – RH	72	No. 8-32 x 1/2 Screw
26	No. 10-32 x 1/4 Screw	63	Door Spring	73	Hinge – LH
27	Timer Knob	64	Door Gasket	74	No. 8-32 x 1/2 Screw
28	Timer Dial	65	Gasket Retainer	75	"O" Ring
29	Timer Skirt Hub	66	No. 8-32 Nut	76	Screw
30	Plug	67	Door Panel	77	Screw
31	No. 10-32 x 5/8 Screw		White	78	Pin
32	Grommet		Pink	79	No. 8-32 x 1/4 Screw
33	No. 10-32 Nut		Yellow	80	Fibre Washer
34	No. 8-32 x 3/8 Screw		Copper	81	Screw
36	Dispenser Solenoid		Brushed Chrome	84	Grille and Screen
37	No. 6-20 x 5/16 Screw		Turquoise	85	Boot
38	Retainer Cover	68	Door Panel Trim	86	Grille
39	Cover Spring		(Used W/Overlay)		
41	No. 8-32 Nut		Door Panel Trim		
43	Linkage		(Flush)		
44	Roll Pin		Door Panel Trim		
45	Lever Arm		(Used W/Wood Overlay)		

TYPICAL OF UNDERCOUNTER DISHWASHERS
Order Parts by Model Number

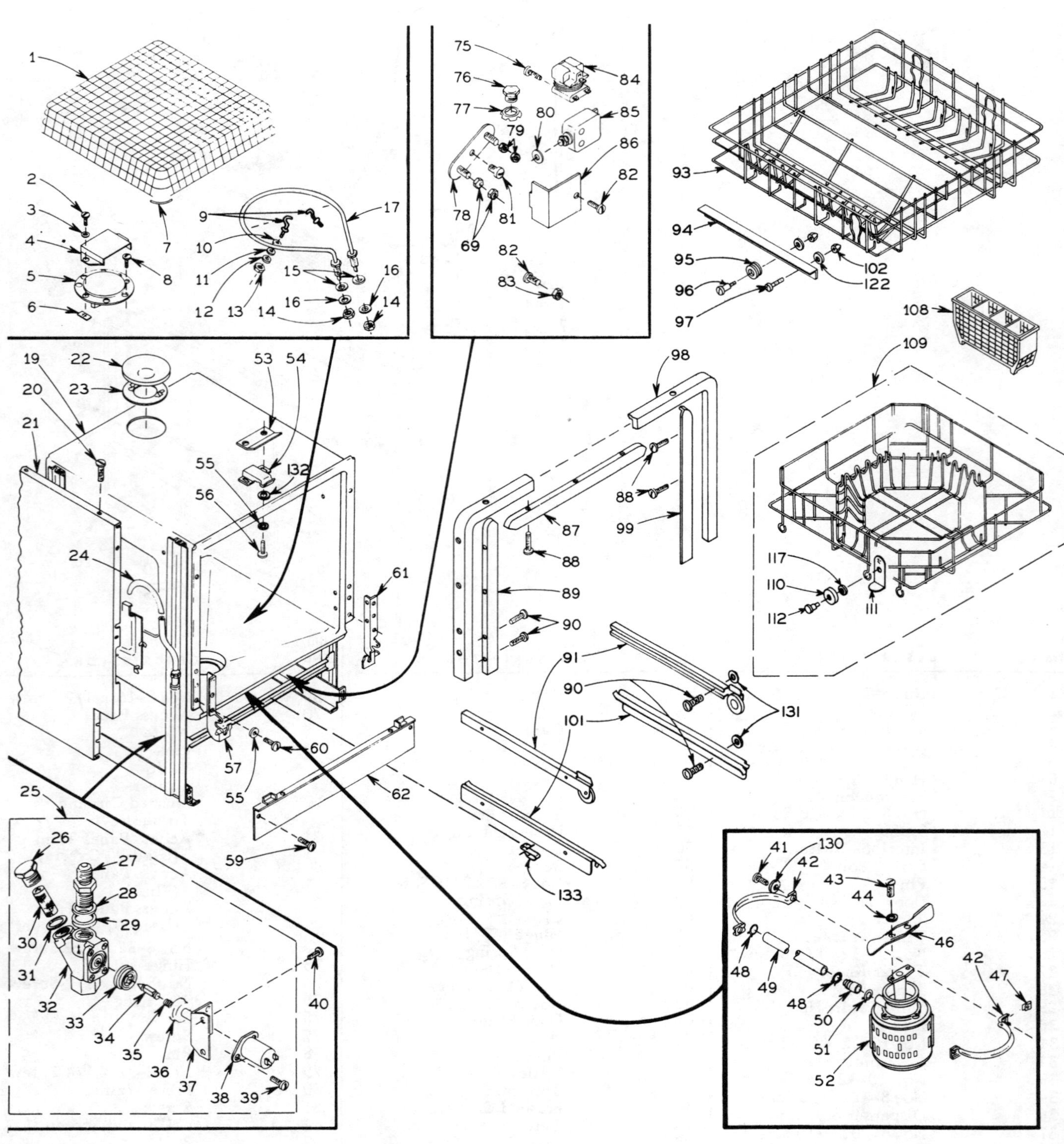

TYPICAL OF UNDERCOUNTER DISHWASHERS
Order Parts by Model Number

Illus. No.	DESCRIPTION	Illus. No.	DESCRIPTION	Illus. No.	DESCRIPTION
1	Impeller Guard	33	Diaphragm	88	No. 10-32 x 1/2 Screw
2	No. 8-32 x 5/8 Screw	34	Armature	89	Gasket Retainer - L.H.
3	Sealing Washer	35	Spring	90	No. 10-32 x 3/8 Screw
4	Protector	36	Guide	91	Upper Track - L.H.
5	Pressure Switch	37	Bracket		Upper Track - R.H.
6	Speed Nut	38	Solenoid	93	Upper Dishrack
7	Corner Guard	39	No. 8-32 Screw	94	Upper Dishrack Slide - L.H.
8	No. 8-32 x 9/16 Screw	40	1/4-20 x 3/8 Screw		Upper Dishrack Slide - R.H.
9	Heater Support	41	No. 10-16 x 1 Screw		
10	Retainer Washer	42	Clamp	95	Wheel
11	Washer	43	No. 12-24 x 1/2 Screw	96	Shoulder Screw
12	Washer	44	Lockwasher	97	No. 8-32 x 7/16 Screw
13	No. 10-32 Nut	46	Impeller	98	Gasket
14	7/16-20 Nut	47	Speed Nut	99	Gasket Retainer - R.H.
15	Fibre Washer	48	Hose Clamp	101	Track - Lower L.H.
16	1/2 Washer	49	Drain Hose		Track - Lower R.H.
17	Heater	50	Adaptor	102	Cap Nut
19	Tub	51	Hose Clamp	108	Silverware Basket
20	No. 8-15 x 1/2 Screw	52	Motor and Pump	109	Lower Dishrack Assembly
21	Side Panel, R.H.	53	Strike Plate	110	Wheel
	White	54	Strike Latch	111	Lower Dishrack Hold Down
	Pink	55	Fibre Washer	112	Rivet
	Yellow	56	No. 8-32 x 1/2 Screw	117	Washer
	Copper	57	Hinge Bracket - L.H.	122	Washer
	Turquoise	59	No. 10-32 x 3/8 Screw	130	1/4 Washer
	Side Panel, L.H.	60	No. 8-32 x 7/16 Screw	131	Washer
	White	61	Hinge Bracket - R.H.	132	No. 8 Lockwasher
	Pink	62	Toe Plate	133	Clip
	Yellow	69	No. 10-32 Nut		
	Copper	75	No. 8-32 x 1/4 Screw		
	Turquoise	76	Bushing		
22	Pressure Relief Valve	77	Nut		
23	Baffle	78	Terminal Block		
24	Water Inlet Tube	79	No. 10-32 Nut		
25	Water Inlet Valve	80	Lockwasher		
26	Plug	81	No. 8-32 x 3/8 Screw		
27	Outlet Fitting	82	No. 8-32 x 3/8 Screw		
28	Gasket	83	No. 8-32 Nut		
29	Flow Washer	84	Relay		
30	Screen	85	Overload Switch		
31	Gasket	86	Cover		
32	Body	87	Gasket Retainer - Top		

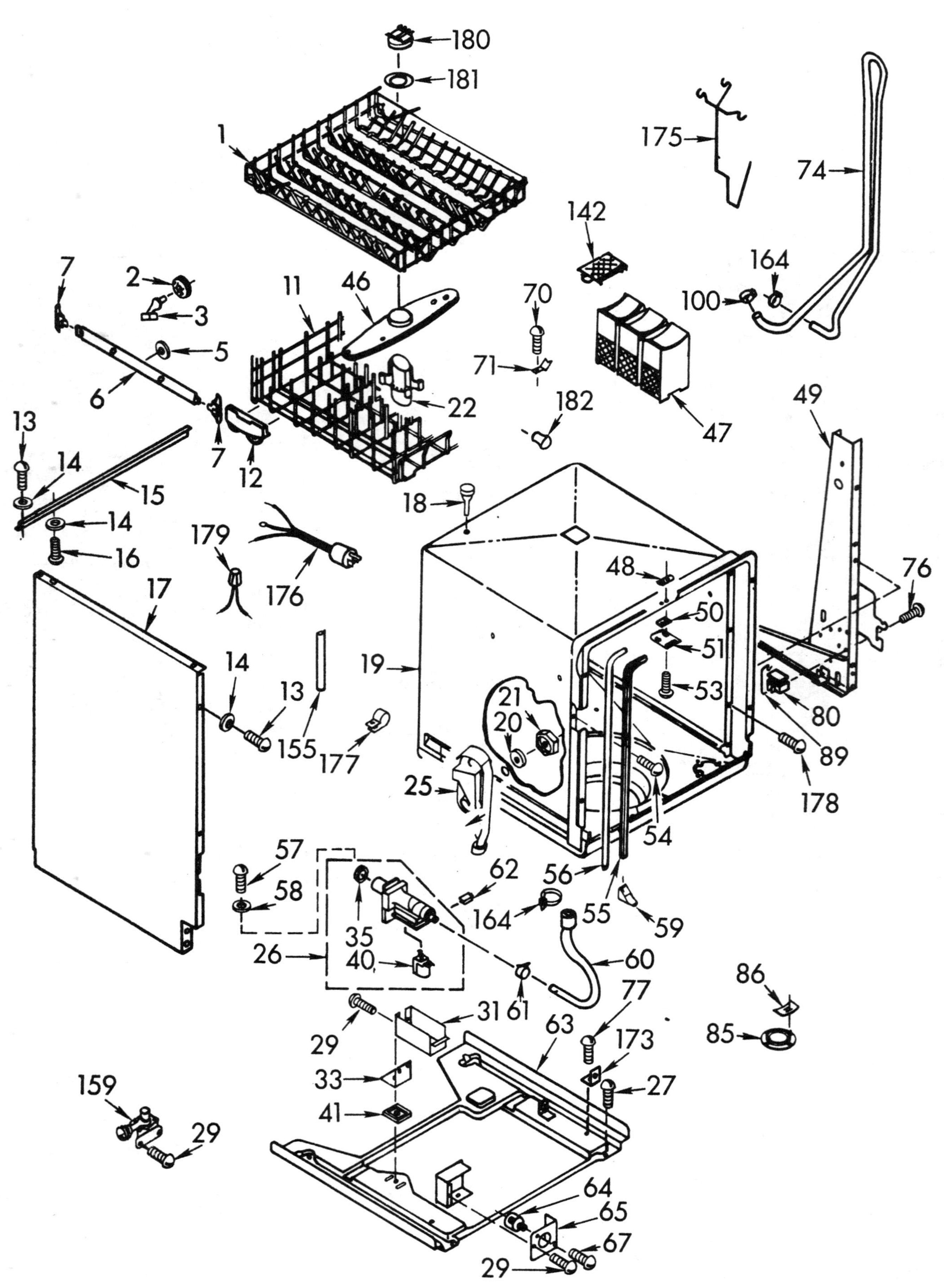
180
181
1
7
2
3
11
46
70
142
164
100
175
74
71
5
6
22
182
47
49
13
14
15
14
16
18
179
176
17
48
50
51
53
76
14
19
21
20
80
155 177
25
89
178
54
57
58
62
56
55
59
35
40
26
164
60
31 61
29
63
77
86
85
33
173
27
159
41
29
64
65
67
29

Illus. No.	DESCRIPTION	Illus. No.	DESCRIPTION	Illus. No.	DESCRIPTION
1	DISHRACK, UPPER	33	ANGLE	100	CLAMP, HOSE
2	WHEEL	35	WASHER, FLOW	142	COVER, BASKET
3	AXLE	40	SOLENOID	155	TUBING
5	WHEEL (4)	41	PAD, MOUNTING	159	VALVE, AIR
6	TRACK (2)	46	ARM, SPRAY (UPPER)	162	SCREW
7	STOP	47	BASKET, SILVERWARE (2)	164	STRAP, HOSE
11	DISHRACK, LOWER	48	PLATE, TAPPING	173	ANGLE, BASE PAN
12	WHEEL (4)	49	LEG & CHANNEL (R.S.)	174	SCREW
13	SCREW, 8-32 x 3/8		LEG & CHANNEL (L.S.)	175	SUPPORT, HOSE
14	WASHER, FIBER	50	SHIM	176	CORD, POWER
15	BRACKET, PANEL	51	STRIKE	177	CLAMP, WIRE
16	SCREW, 8-32 x 1/2	53	SCREW, 8-32 x 1/2	178	SCREW
17	PANEL, R.H.	54	SCREW	179	CONNECTOR
	WHITE	55	RETAINER (L.S.)	180	SUPPORT, SPRAY ARM
	SHADED COPPER		RETAINER (R.S.)	181	RING, SUPPORT SPRAY ARM
	SHADED AVOCADO	56	GASKET, TUB		
	SHADED SAPPHIRE BLUE	57	SCREW	182	BUMPER
	SHADED FAWN	58	WASHER, METAL		
	HARVEST GOLD	59	GASKET (L.S.)		
	PANEL, L.H.		GASKET (R.S.)		
	WHITE	60	HOSE, INLET		
	SHADED COPPER	61	CLAMP, HOSE		
	SHADED FAWN	62	STRAP, GROUND		
	SHADED AVOCADO	63	PAN, BASE		
	SHADED SAPPHIRE BLUE	64	RECEPTACLE		
	HARVEST GOLD	65	COVER, TERMINAL		
18	BUMPER, STEM	67	SCREW, 6-32 x 3/8		
19	TUB	70	SCREW, 6-32 x 1/2		
20	GASKET	71	CLIP, GROUNDING		
21	NUT	72	CLAMP, SPEED		
22	TOWER	74	HOSE, DRAIN		
25	TUBE, INLET	76	SCREW		
26	WATER INLET	77	SCREW, 10-12 x 1/2		
	VALVE ASSEMBLY	80	RELAY, STARTING		
27	SCREW	85	SWITCH, PRESSURE		
29	SCREW	86	NUT, SPEED		
31	SHIELD, COIL	89	PROTECTOR, RELAY		

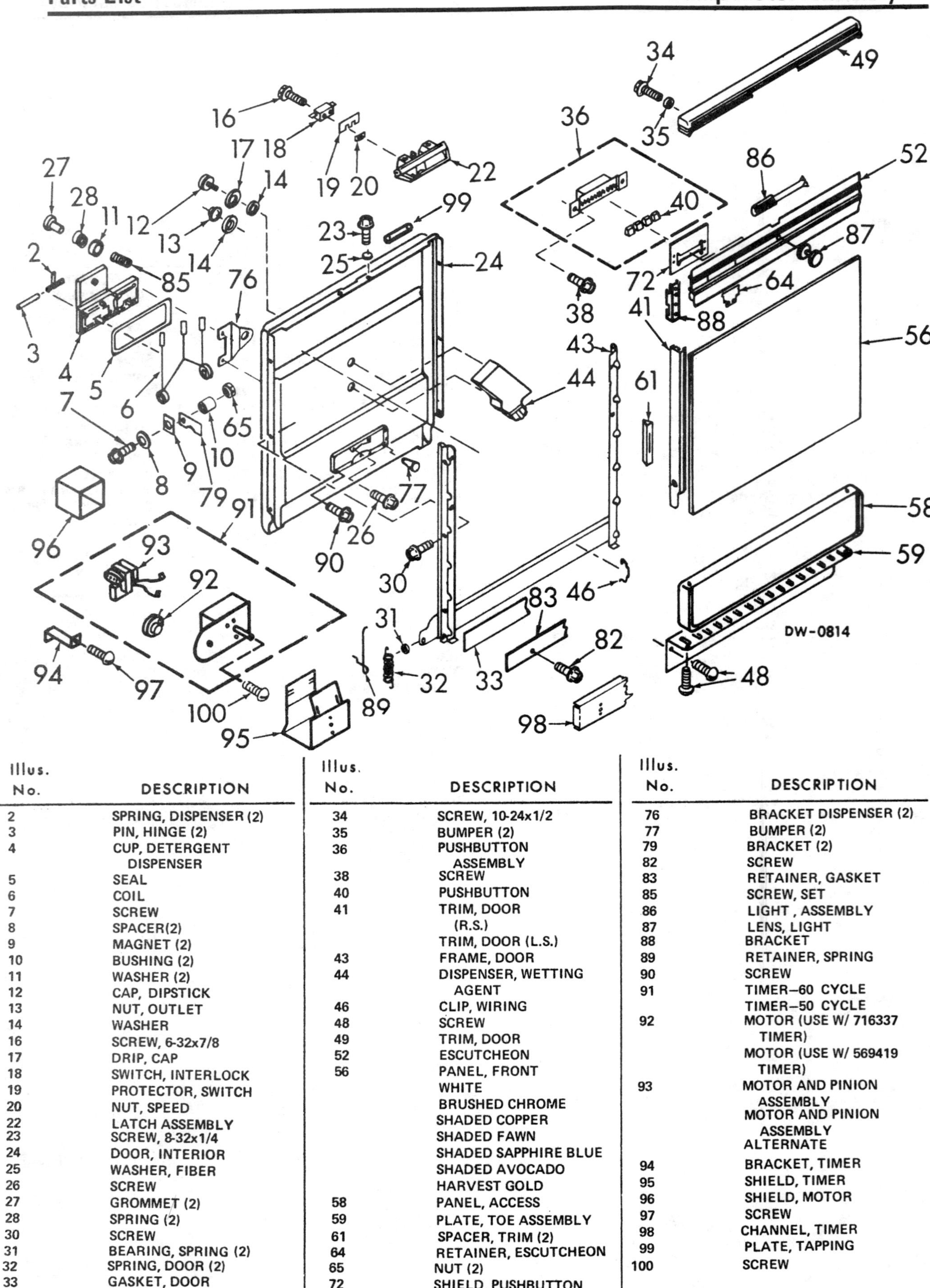

Illus. No.	DESCRIPTION	Illus. No.	DESCRIPTION	Illus. No.	DESCRIPTION
2	SPRING, DISPENSER (2)	34	SCREW, 10-24x1/2	76	BRACKET DISPENSER (2)
3	PIN, HINGE (2)	35	BUMPER (2)	77	BUMPER (2)
4	CUP, DETERGENT DISPENSER	36	PUSHBUTTON ASSEMBLY	79	BRACKET (2)
5	SEAL	38	SCREW	82	SCREW
6	COIL	40	PUSHBUTTON	83	RETAINER, GASKET
7	SCREW	41	TRIM, DOOR (R.S.)	85	SCREW, SET
8	SPACER (2)		TRIM, DOOR (L.S.)	86	LIGHT , ASSEMBLY
9	MAGNET (2)	43	FRAME, DOOR	87	LENS, LIGHT
10	BUSHING (2)	44	DISPENSER, WETTING AGENT	88	BRACKET
11	WASHER (2)	46	CLIP, WIRING	89	RETAINER, SPRING
12	CAP, DIPSTICK	48	SCREW	90	SCREW
13	NUT, OUTLET	49	TRIM, DOOR	91	TIMER—60 CYCLE
14	WASHER	52	ESCUTCHEON		TIMER—50 CYCLE
16	SCREW, 6-32x7/8	56	PANEL, FRONT	92	MOTOR (USE W/ 716337 TIMER)
17	DRIP, CAP		WHITE		MOTOR (USE W/ 569419 TIMER)
18	SWITCH, INTERLOCK		BRUSHED CHROME	93	MOTOR AND PINION ASSEMBLY
19	PROTECTOR, SWITCH		SHADED COPPER		MOTOR AND PINION ASSEMBLY ALTERNATE
20	NUT, SPEED		SHADED FAWN		
22	LATCH ASSEMBLY		SHADED SAPPHIRE BLUE		
23	SCREW, 8-32x1/4		SHADED AVOCADO	94	BRACKET, TIMER
24	DOOR, INTERIOR		HARVEST GOLD	95	SHIELD, TIMER
25	WASHER, FIBER	58	PANEL, ACCESS	96	SHIELD, MOTOR
26	SCREW	59	PLATE, TOE ASSEMBLY	97	SCREW
27	GROMMET (2)	61	SPACER, TRIM (2)	98	CHANNEL, TIMER
28	SPRING (2)	64	RETAINER, ESCUTCHEON	99	PLATE, TAPPING
30	SCREW	65	NUT (2)	100	SCREW
31	BEARING, SPRING (2)	72	SHIELD, PUSHBUTTON		
32	SPRING, DOOR (2)				
33	GASKET, DOOR FRAME				

TYPICAL OF UNDERCOUNTER DISHWASHERS
Order Parts by Model Number

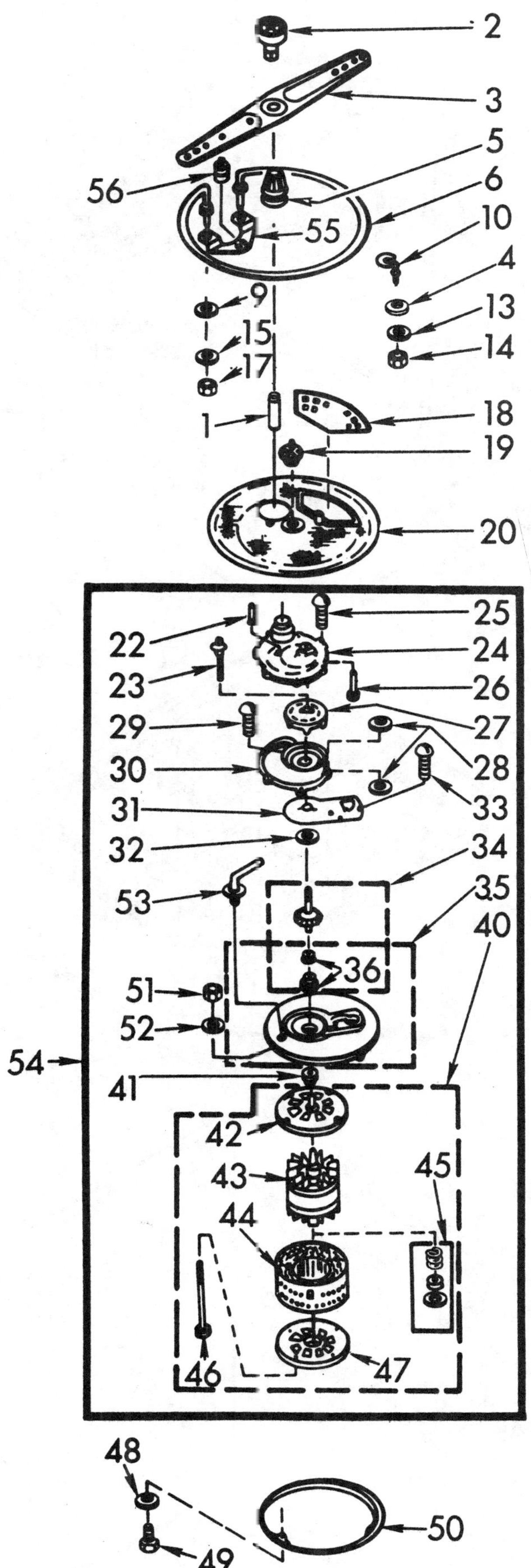

Illus. No.	DESCRIPTION
1	NOZZLE, JET
2	CAP, NOZZLE
3	ARM, SPRAY (LONER)
4	WASHER
5	HUB
6	HEATER
9	LOCKWASHER
10	SUPPORT, HEATER (2)
13	WASHER
14	NUT
15	WASHER, METAL
17	NUT, 7/16—20
18	GUARD, SUMP SCREEN
19	NUT, SUMP SCREEN
20	SCREEN, SUMP
22	PIPE, STAND
23	SCREW ASSEMBLY IMPELLER (10-32 x 2")
24	COVER, VOLUTE
25	SCREW, 8-18 x 1"
26	TURE, VENT
27	IMPELLER, WASH
28	WASHER, VENT
29	SCREW, 8-32 x 1-1/8
30	BASE, VOLUTE
31	COVER, DRAIN PUMP
32	"O" RING
33	SCREW, 8-32 x 1/2
34	IMPELLER (ALUMINIUM DIE CAST)
	IMPELLER (STAINLESS STEEL)
35	PLATE, PUMP MOUNTING
36	SEAL
40	MOTOR
41	SPACER, IMPELLER
42	FRAME, PUMP SIDE
43	ROTOR
44	STATOR — 60 CYCLE
	STATOR — 50 CYCLE
45	THRUST, SPRING AND WASHER
46	SCREW
47	FRAME, OUTER
48	WASHER, RUBBER
49	SCREW
50	GASKET, PUMP
51	NUT
52	WASHER
53	TUB BLEEDER
54	PUMP & MOTOR ASSEMBLY
55	BRACKET, THERMOSTAT
56	THERMOSTAT ASSEMBLY

TYPICAL OF UNDERCOUNTER DISHWASHERS
Order Parts by Model Number

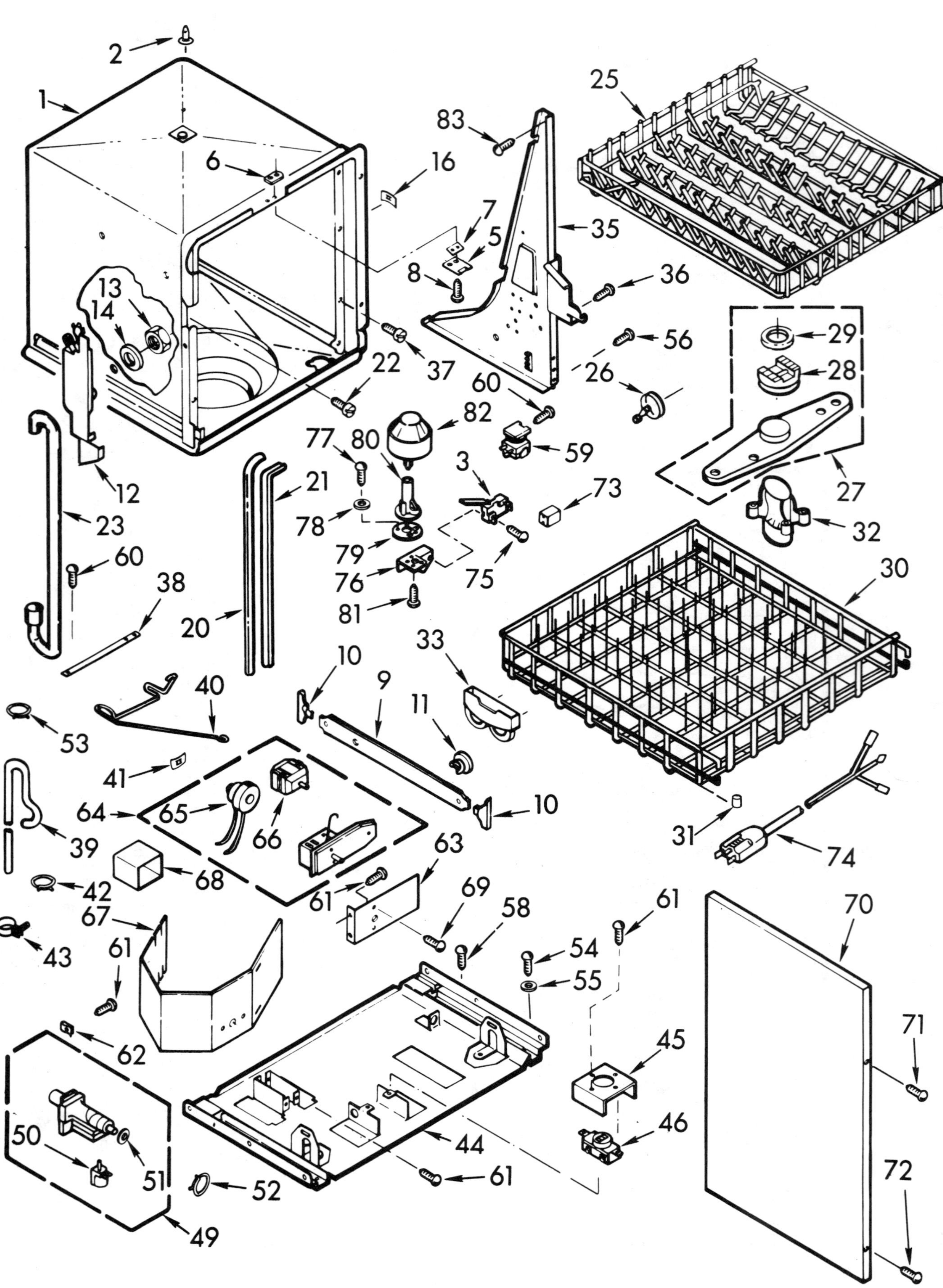

25
2
1
6
16
83
7
5
8
35
13
14
36
22
37
60
56
26
82
59
77
80
3
73
29
28
78
79
76
81
75
27
32
21
12
23
60
38
20
10
33
30
9
11
53
40
41
64
65
66
63
10
31
74
39
68
61
69
58
61
70
42
67
61
54
55
43
45
50
62
44
46
71
51
52
61
49
72

Illus. No.	DESCRIPTION	Illus. No.	DESCRIPTION	Illus. No.	DESCRIPTION
1	TUB ASSEMBLY	35	LEG & HINGE ASSEMBLY L.H.	65	MOTOR 60 CYCLE
2	BUMPER		LEG & HINGE ASSEMBLY R.H.		MOTOR 60 CYCLE
3	SWITCH, PRESSURE	36	SCREW		(ALTERNATE)
5	STRIKE	37	SCREW	66	MOTOR & PINION
6	PLATE, TAPPING	38	PLATE, TAPPING		ASSEMBLY (ALT.)
7	SHIM	39	HOSE, DRAIN		MOTOR & PINION
8	SCREW	40	SUPPORT		ASSEMBLY (ALT.)
9	TRACK (2)	41	CLIP STUD	67	BARRIER
10	STOP(4)	42	CLAMP HOSE DRAIN	68	SHIELD
11	WHEEL ASSEMBLY (4)	43	STRAP, HOSE	69	SCREW
12	TUBE, INLET	44	PAN, BASE	70	PANEL R.S.
13	NUT	45	COVER, TERMINAL BOX		WHITE
14	GASKET	46	RECEPTACLE		COPPERTONE
16	NUT, SPRING	49	VALVE, WATER INLET		AVOCADO
20	GASKET	50	COIL		HARVEST GOLD
21	RETAINER R.S.	51	WASHER, FLOW		PANEL L.S.
	RETAINER L.S.	52	CLAMP		WHITE
22	SCREW	53	CLAMP		COPPERTONE
23	HOSE, INLET	54	SCREW		AVOCADO
25	DISHRACK, UPPER	55	WASHER		HARVEST GOLD
26	WHEEL & AXLE ASSEMBLY	56	SCREW	71	SCREW
27	ARM, SPRAY	58	SCREW	72	SCREW
28	SUPPORT, SPRAY ARM	59	RELAY, STARTING	73	BARRIER, RELAY
29	RING, SUPPORT	60	SCREW	74	CORD, POWER
	SPRAY ARM	61	SCREW, GROUND	75	SCREW
30	DISHRACK, LOWER	62	CLAMP	76	BRACKET, OVERFLOW
	(INCLUDES ILLUS. 31)	63	BRACKET	77	SCREW
31	BUMPER	64	TIMER 60 CYCLE	78	WASHER
32	TOWER			79	GASKET, OVERFLOW
33	WHEEL & SHROUD			80	STANDPIPE, OVERFLOW
	ASSEMBLY			81	SCREW
				82	FLOAT
				83	SCREW
				84	SCREW, GROUND

TYPICAL OF UNDERCOUNTER DISHWASHERS
Order Parts by Model Number

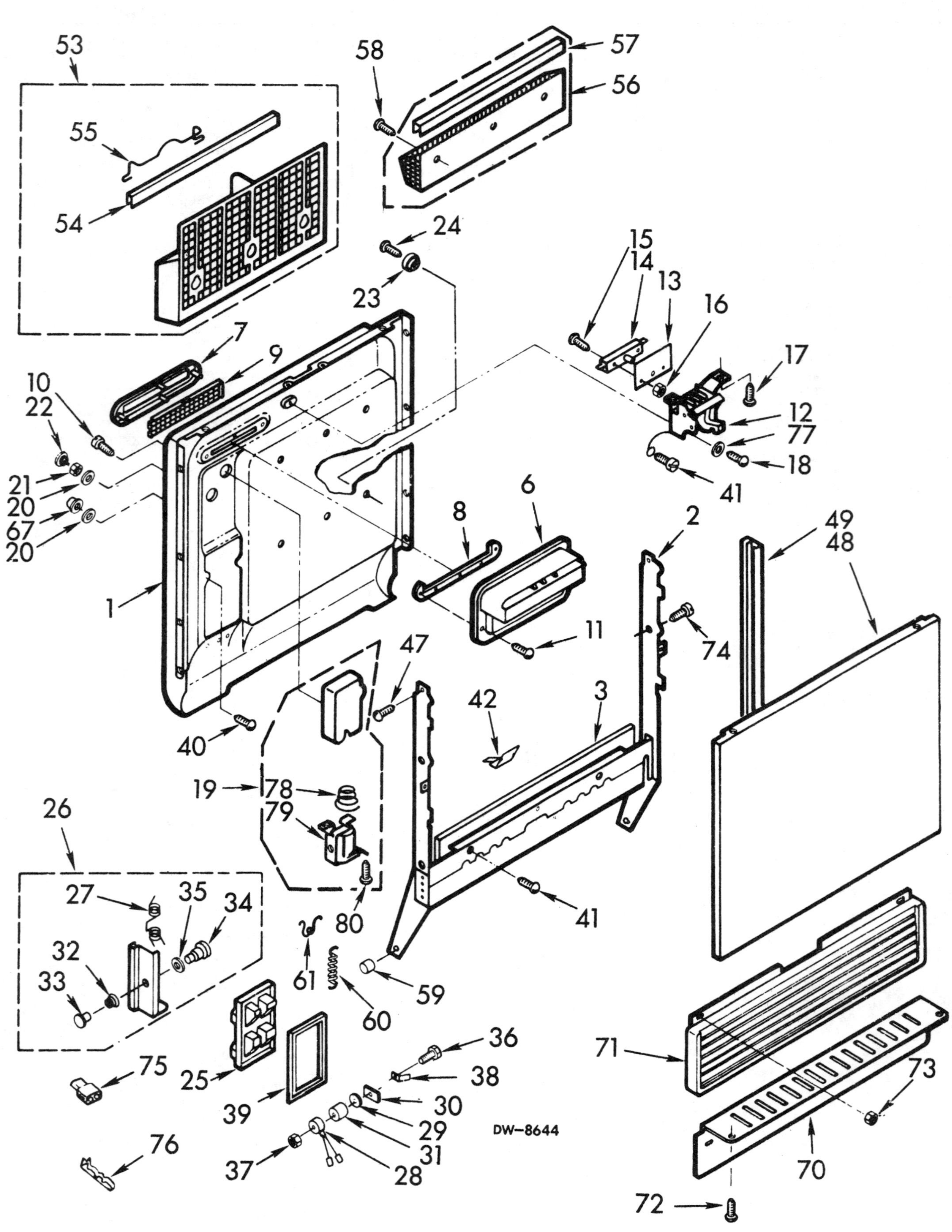

TYPICAL OF UNDERCOUNTER DISHWASHERS
Order Parts by Model Number

Illus. No.	DESCRIPTION	Illus. No.	DESCRIPTION	Illus. No.	DESCRIPTION
1	DOOR, INNER	27	SPRING, LID	57	TRIM BASKET
2	DOOR FRAME	28	COIL ASSEMBLY	58	SCREW
	(INCLUDES ILLUS. 3)	29	SPACER, MAGNET	59	BEARING, SPRING
3	GASKET, DOOR	30	MAGNET	60	SPRING, DOOR
6	BODY, VENT	31	BUSHING	61	RETAINER, SPRING
7	BEZEL, VENT	32	SPRING	67	DRIP CAP
8	GASKET	33	COVER, BUTTON	70	PLATE, TOE
9	SCREEN	34	PIN BOTTOM	71	PANEL, ACCESS
10	SCREW	35	WASHER	72	SCREW, 8-32 x 3/8
11	SCREW	36	SCREW, 4-40 x 7/8	73	NUT, HEX 8-32
12	LATCH ASSEMBLY	37	NUT	74	SCREW
13	PROTECTOR, SWITCH	38	TERMINAL	75	PLUG
14	SWITCH, INTERLOCK	39	GASKET DETERGENT	76	TERMINAL
15	SCREW		DISPENSER	77	WASHER, SEAL
16	NUT	40	SCREW, 8 x 1/2	78	SPRING
17	SCREW	41	SCREW	79	COIL ASSEMBLY
18	SCREW	42	CLIP, GROUND	80	SCREW, 8 x 5/8 (SELF
19	DISPENSER, WETTING	47	SCREW		TAPPING)
	AGENT ASSEMBLY	48	PANEL FRONT		
20	WASHER		WHITE		FOLLOWING PARTS NOT ILLUSTRATED
21	NUT		AVOCADO		HARNESS, WIRING
22	CAP ASSEMBLY		COPPERTONE		INSTRUCTIONS,
23	BUTTON, RETAINER		HARVEST GOLD		INSTALLATION
24	SCREW		BRUSH CHROME		USE & CARE GUIDE
25	BODY, DETERGENT	49	TRIM, DOOR R.H.		LIST, PARTS
	DISPENSER		TRIM, DOOR L.H.		
26	LID, DETERGENT	53	BASKET, SILVERWARE		
	DISPENSER ASSEMBLY		ASSEMBLY		
		54	TRIM, BASKET		
		55	HANDLE, BASKET		
		56	UTENSIL BASKET		
			ASSEMBLY		

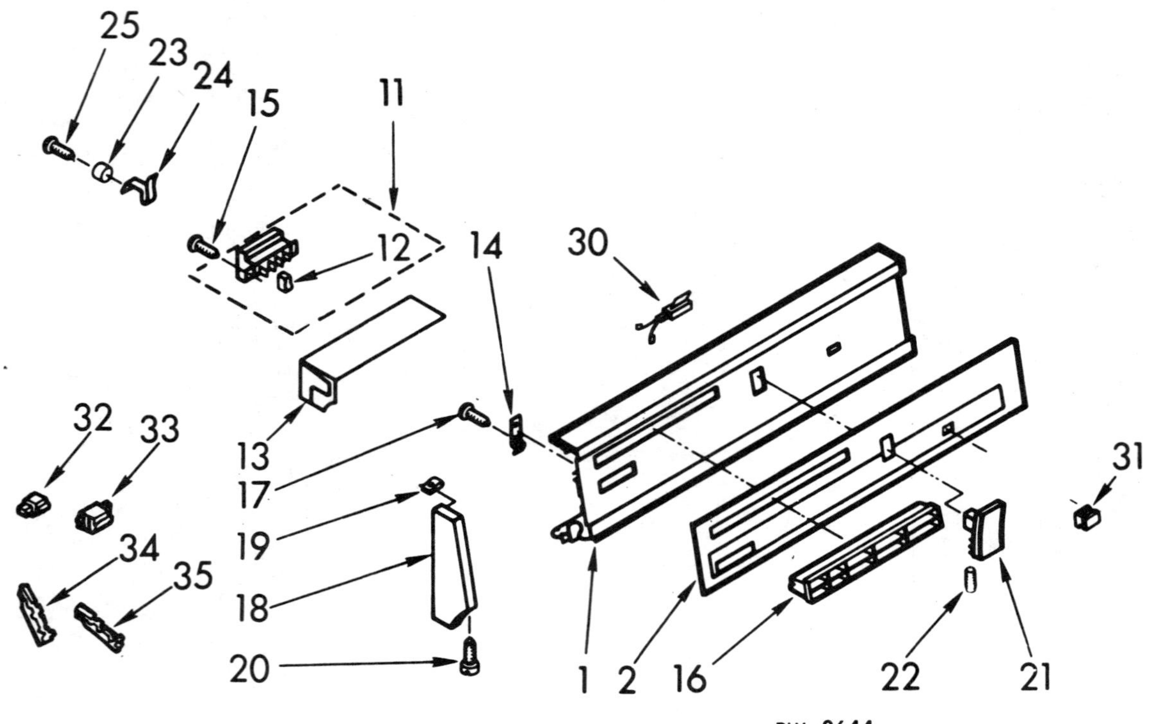

Illus. No.	DESCRIPTION	Illus. No.	DESCRIPTION	Illus. No.	DESCRIPTION
1	CONSOLE, FRONT	18	CAP, END R.H.	25	SCREW
2	PANEL, ESCUTCHEON		CAP, END L.H.	30	LIGHT ASSEMBLY
11	SWITCH, PUSHBUTTON	19	CLIP	31	LENS
12	BUTTON	20	SCREW	32	CONNECTOR, TERMINAL
13	BARRIER, PUSHBUTTON	21	HANDLE ASSEMBLY		MOTOR
14	BRACKET, SWITCH		(INCLUDES ILLUS. 22)	33	PLUG, TERMINAL
15	SCREW	22	SCREW, SET		
16	GRILL, VENT	23	BUMPER	34	TERMINAL
17	SCREW	24	CLIP	35	TERMINAL

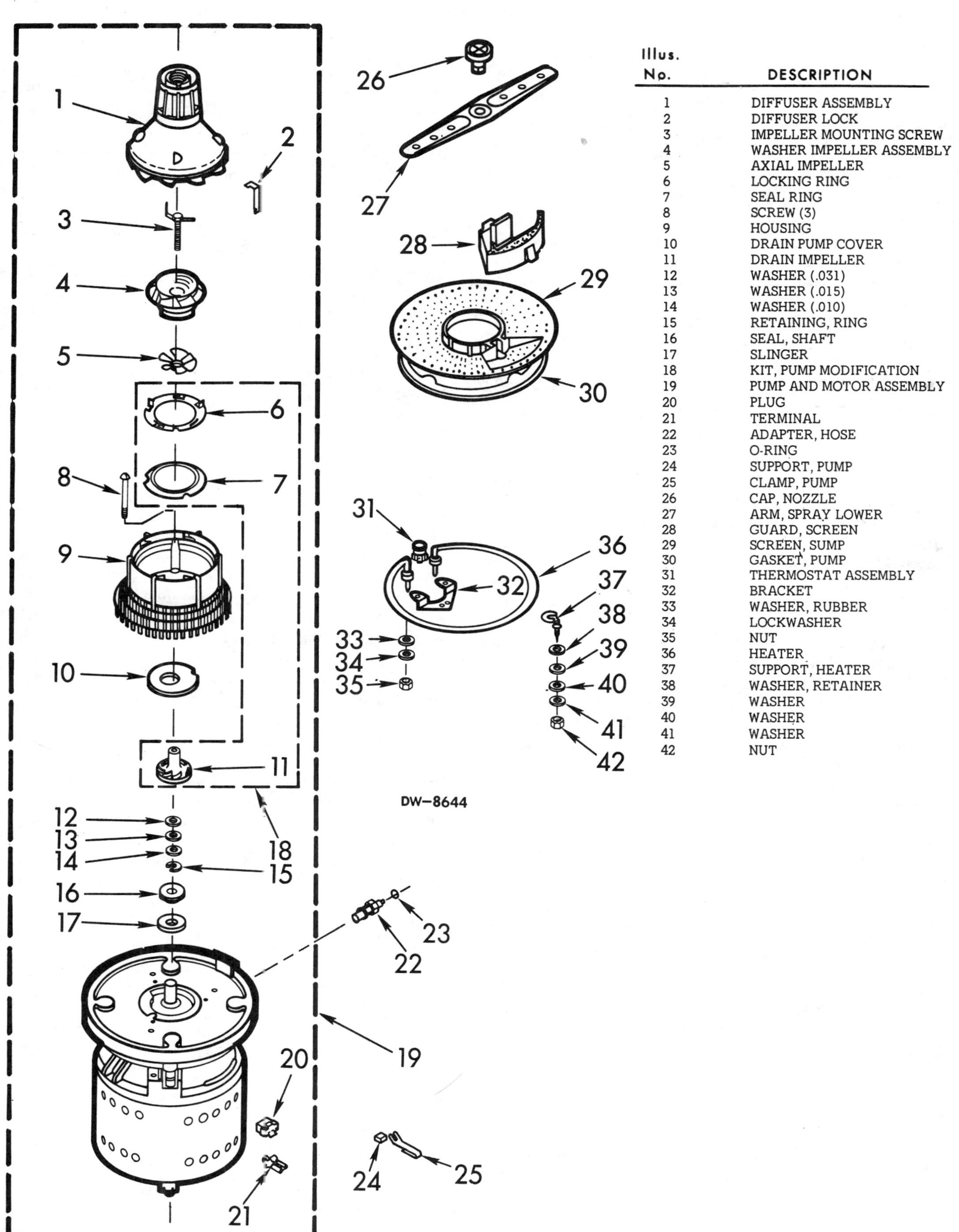

Illus. No.	DESCRIPTION
1	DIFFUSER ASSEMBLY
2	DIFFUSER LOCK
3	IMPELLER MOUNTING SCREW
4	WASHER IMPELLER ASSEMBLY
5	AXIAL IMPELLER
6	LOCKING RING
7	SEAL RING
8	SCREW (3)
9	HOUSING
10	DRAIN PUMP COVER
11	DRAIN IMPELLER
12	WASHER (.031)
13	WASHER (.015)
14	WASHER (.010)
15	RETAINING, RING
16	SEAL, SHAFT
17	SLINGER
18	KIT, PUMP MODIFICATION
19	PUMP AND MOTOR ASSEMBLY
20	PLUG
21	TERMINAL
22	ADAPTER, HOSE
23	O-RING
24	SUPPORT, PUMP
25	CLAMP, PUMP
26	CAP, NOZZLE
27	ARM, SPRAY LOWER
28	GUARD, SCREEN
29	SCREEN, SUMP
30	GASKET, PUMP
31	THERMOSTAT ASSEMBLY
32	BRACKET
33	WASHER, RUBBER
34	LOCKWASHER
35	NUT
36	HEATER
37	SUPPORT, HEATER
38	WASHER, RETAINER
39	WASHER
40	WASHER
41	WASHER
42	NUT

INDEX

INDEX

GEM PRODUCTS, INC. MASTER PUBLICATIONS

APPLIANCE REPAIR MANUALS

REPAIR-MASTER® For Automatic Washers, Dryers & Dishwashers

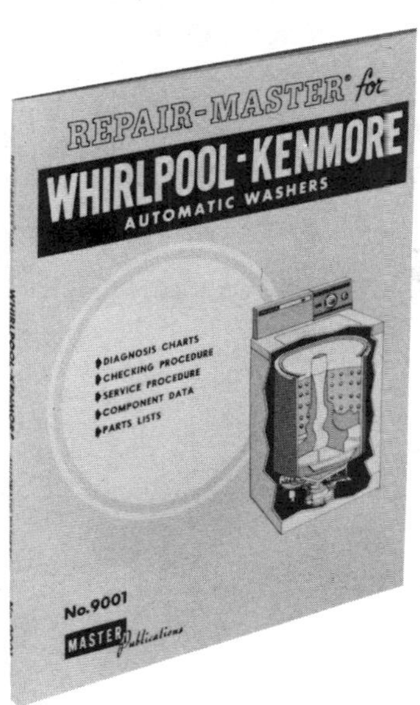

These Repair-Masters® offer a quick and handy reference for the diagnosis and correction of service problems encountered on home appliances. They contain many representative illustrations, diagrams and photographs to clearly show the various components and their service procedure.

Diagnosis and repair charts provide step-by-step detailed procedures and instruction to solve the most intricate problems encountered in the repair of washers, dryers and dishwashers.

These problems range from timer calibrations to complete transmission repair, all of which are defined and explained in easy to read terms.

The Repair-Masters® are continually updated to note the latest changes or modifications in design or original parts. These changes and modifications are explained in their service context to keep the serviceman abreast of the latest developments in the industry.

AUTOMATIC WASHERS
9001	Whirlpool
9003	General Electric
9005	Westinghouse Front Loading
9009	Frigidaire
9010	Maytag
9011	Philco-Bendix Front Loading
9012	Speed Queen
9015	Norge – Plus Capacity
9016	Frigidaire Roller-Matic
9017	Westinghouse Top Loading

DISHWASHERS
5552	General Electric
5553	Kitchenaid
5554	Westinghouse
5555	D & M
5556	Whirlpool
5557	Frigidare

CLOTHES DRYERS
8051	Whirlpool-Kenmore
8052	General Electric
8053	Hamilton
8055	Maytag
8056	Westinghouse
8057	Speed Queen
8058	Franklin
8059	Frigidaire

***The D & M Dishwasher Repair-Masters covers the following brands: Admiral — Caloric — Chambers — Frigidaire (Portable) — Gaffers and Sattler — Gibson — Kelvinator — Kenmore — Magic Chef — Magic Maid — Norge — Philco — Pioneer — Preway — Roper — Wedgewood — Westinghouse (Portable)**

The Repair-Masters® series will take the guesswork out of service repairs.

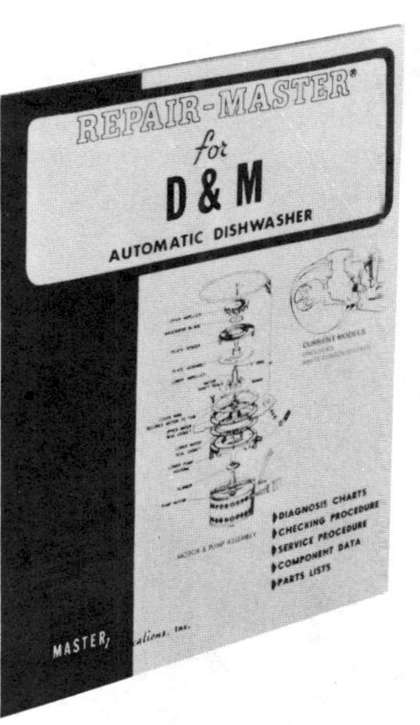

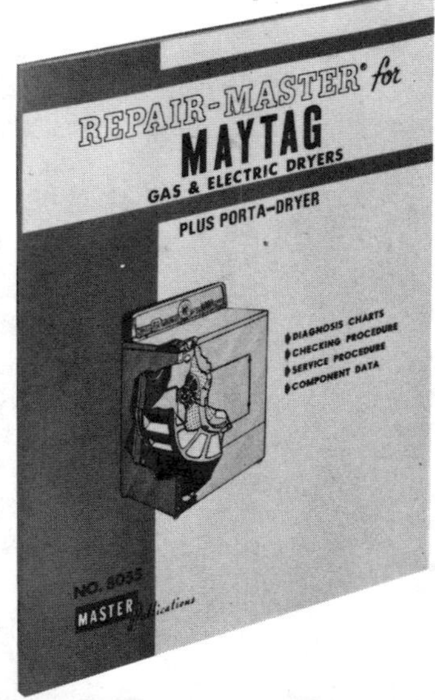

REPAIR-MASTER® For Window Air Conditioners

This Repair-Master® offers a quick and handy reference for the diagnosis and correction of service problems. Many illustrations, diagrams and photographs are included to clearly show the various components and their service procedures.

A Complete Service and Repair Guide!

PRINCIPLES OF REFRIGERATION

- DIAGNOSIS CHARTS
- TEST CORD TESTING
- PROCEDURES
- STEP-BY-STEP REPAIR
- SHOP PROCEDURES

7541
84 Pages

REPAIR MASTER® For Domestic Automatic Icemakers

Includes ten different designs of Automatic Ice Makers presently found in domestic refrigerators.

A special section is included for Whirlpool design self-contained Ice Maker. Illustrations, charts and new parts design information makes this Repair-Master® an important tool.

- DIAGNOSIS CHARTS
- TESTING PROCEDURES
- ADJUSTMENT PROCEDURES
- STEP-BY-STEP REPAIRS

7531
176 Pages

TECH-MASTER® For Refrigerators & Freezers

For All Major Brands

Offering a quick reference by brand, year, and model number.

Handy 8½" x 5½" book is designed to be easily carried in the tool box or service truck. The first time you use these books they will more than repay the nominal purchase price.

Operating Information for Over 15,000 Models

- TEMPERATURE CONTROL PART NUMBERS
- CUT-IN AND CUT-OUT SETTINGS
- UNIT HORSEPOWER
- REFRIGERANT AND OIL CHARGE IN OUNCES
- SUCTION AND HEAD PRESSURES
- START AND RUN WATTAGES
- COMPRESSOR TERMINAL HOOK-UPS
- DEFROST HEATER WATTAGES

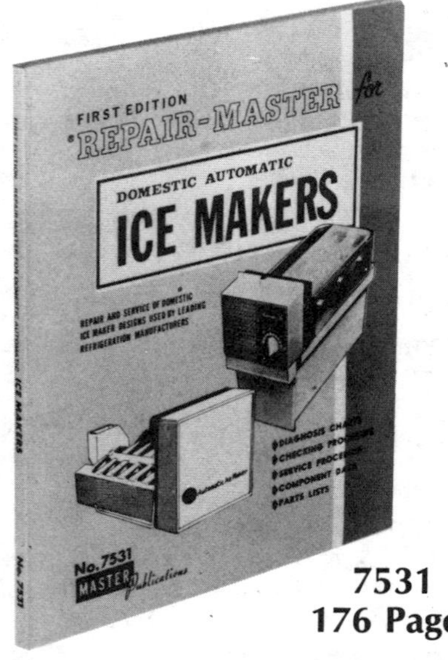

No. 7550	No. 7550-2	No. 7550-3	No. 7550-4
1950-1965	1966-1970	1971-1975	1976-1980
429 Pages	377 Pages	332 Pages	378 Pages